NATURAL POLYMERS FOR PHARMACEUTICAL APPLICATIONS

Volume 2: Marine- and Microbiologically Derived Polymers

Natural Polymers for Pharmaceutical Applications, 3-volume set:

Natural Polymers for Pharmaceutical Applications
Volume 1: Plant-Derived Polymers

Natural Polymers for Pharmaceutical Applications
Volume 2: Marine- and Microbiologically Derived Polymers

Natural Polymers for Pharmaceutical Applications
Volume 3: Animal-Derived Polymers

NATURAL POLYMERS FOR PHARMACEUTICAL APPLICATIONS

Volume 2: Marine- and Microbiologically Derived Polymers

Edited by

Amit Kumar Nayak, PhD
Md Saquib Hasnain, PhD
Dilipkumar Pal, PhD

AAP APPLE ACADEMIC PRESS

Apple Academic Press Inc. | Apple Academic Press Inc.
3333 Mistwell Crescent | 1265 Goldenrod Circle NE
Oakville, ON L6L 0A2 | Palm Bay, Florida 32905
Canada | USA

© 2020 by Apple Academic Press, Inc.

First issued in paperback 2021

Exclusive worldwide distribution by CRC Press, a member of Taylor & Francis Group
No claim to original U.S. Government works

Natural Polymers for Pharmaceutical Applications, Volume 2:
Marine- and Microbiologically Derived Polymers
ISBN 13: 978-1-77463-184-3 (pbk)
ISBN 13: 978-1-77188-846-2 (hbk)

Natural Polymers for Pharmaceutical Applications, 3-volume set
ISBN 13: 978-1-77188-846-2 (hbk)
ISBN 13: 978-0-42932-829-9 (ebk)

Library and Archives Canada Cataloguing in Publication

Title: Natural polymers for pharmaceutical applications / edited by Amit Kumar Nayak, Md Saquib Hasnain, Dilipkumar Pal.

Names: Nayak, Amit Kumar, 1979- editor. | Hasnain, Md Saquib, 1984- editor. | Pal, Dilipkumar, 1971- editor.

Description: Includes bibliographical references and indexes.

Identifiers: Canadiana (print) 20190146095 | Canadiana (ebook) 20190146117 | ISBN 9781771888448 (set ; hardcover) | ISBN 9781771888455 (v. 1 ; hardcover) | ISBN 9781771888462 (v. 2; hardcover) | ISBN 9781771888479 (v. 3 ; hardcover) | ISBN 9780429328121 (set ; ebook) | ISBN 9780429328251 (v. 1 ; ebook) | ISBN 9780429328299 (v. 2 ; ebook) | ISBN 9780429328350 (v. 3; ebook)

Subjects: LCSH: Polymers in medicine. | LCSH: Biopolymers. | LCSH: Pharmaceutical technology.

Classification: LCC R857.P6 N38 2020 | DDC 615.1/9—dc23

CIP data on file with US Library of Congress

Apple Academic Press also publishes its books in a variety of electronic formats. Some content that appears in print may not be available in electronic format. For information about Apple Academic Press products, visit our website at **www.appleacademicpress.com** and the CRC Press website at **www.crcpress.com**

About the Editors

Amit Kumar Nayak, PhD

Amit Kumar Nayak, PhD, is currently working as an Associate Professor at Seemanta Institute of Pharmaceutical Sciences, Odisha, India. He has earned his PhD in Pharmaceutical Sciences from IFTM University, Moradabad, U.P., India. He has over 10 years of research experience in the field of pharmaceutics, especially in the development and characterization of polymeric composites, hydrogels, novel, and nanostructured drug delivery systems. Till date, he has authored over 120 publications in various high impact peer-reviewed journals and 34 book chapters to his credit. Overall, he has earned highly impressive publishing and cited record in Google Scholar (H-Index: 32, i10-Index: 80). He has been the permanent reviewer of many international journals of high repute. He also has participated and presented his research work at several conferences in India and is a life member of Association of Pharmaceutical Teachers of India (APTI).

Md Saquib Hasnain, PhD

Dr. Md Saquib Hasnain has over 6 years of research experience in the field of drug delivery and pharmaceutical formulation analyses, especially systematic development and characterization of diverse nanostructured drug delivery systems, controlled release drug delivery systems, bioenhanced drug delivery systems, nanomaterials and nanocomposites employing Quality by Design approaches as well as development and characterization of polymeric composites, formulation characterization and many more. Till date he has authored over 30 publications in various high impact peer-reviewed journals, 35 book chapters, 3 books and 1 Indian patent application to his credit. He is also serving as the reviewer of several prestigious journals. Overall, he has earned highly impressive publishing and cited record in Google Scholar (H-Index: 14). He has also participated and presented his research work at over 10 conferences in India, and abroad. He is also the member of scientific societies, i.e., Royal Society of Chemistry, Great Britain, International Association of Environmental and Analytical Chemistry, Switzerland and Swiss Chemical Society, Switzerland.

Dilipkumar Pal, PhD

Dilipkumar Pal, PhD, MPharm, Chartered Chemist, Post Doct (Australia) is an Associate Professor in the Department of Pharmaceutical Sciences, Guru Ghasidash Vishwavidyalaya (A Central University), Bilaspur, C.G., India. He received his master and PhD degree from Jadavpur University, Kolkata and performed postdoctoral research as "Endeavor Post-Doctoral Research Fellow" in University of Sydney, Australia. His areas of research interest include "Isolation, structure Elucidation and pharmacological evaluation of indigenous plants" and "natural biopolymers." He has published 164 full research papers in peer-reviewed reputed national and international scientific journals, having good impact factor and contributed 113 abstracts in different national and international conferences. He has written 1 book and 26 book chapters published by reputed international publishers. His research publications have acquired a highly remarkable cited record in Scopus and Google Scholar (H-Index: 35; i-10-index 82, total citations 3664 till date). Dr. Pal, in his 20 years research-oriented teaching profession, received 13 prestigious national and international professional awards also. He has guided 7 PhD and 39 master students for their dissertation thesis. He is the reviewer and Editorial Board member of 27 and 29 scientific journals, respectively. Dr. Pal has been working as the Editor-in-Chief of one good research journal also. He is the member and life member of 15 professional organizations.

Contents

Contributors ix

Abbreviations xi

Preface xiii

1. **Marine-Derived Polysaccharides: Pharmaceutical Applications** 1
    Dilipkumar Pal, Supriyo Saha, Amit Kumar Nayak, and Md Saquib Hasnain

2. **Pharmaceutical Applications of Alginates** 37
    Amit Kumar Nayak, Moumita Ghosh Laskar, Mohammad Tabish,
    Md Saquib Hasnain, and Dilipkumar Pal

3. **Pharmaceutical Applications of Agar-Agar** 71
    Md Shahruzzaman, Shanta Biswas, Md Nurus Sakib, Papia Haque,
    Mohammed Mizanur Rahman, and Abul K. Mallik

4. **Pharmaceutical Applications of Gellan Gum** 87
    Shakti Nagpal, Sunil Kumar Dubey, Vamshi Krishna Rapalli, and Gautam Singhvi

5. **Pharmaceutical Applications of Carrageenan** 111
    A. Papagiannopoulos and S. Pispas

6. **Pharmaceutical Application of Chitosan Derivatives** 141
    Fiona Concy Rodrigues, Krizma Singh, and Goutam Thakur

7. **Pharmaceutical Applications of Xanthan Gum** 165
    Subham Banerjee and Santanu Kaity

Index 193

Contributors

Subham Banerjee
Department of Pharmaceutics, National Institute of Pharmaceutical Education and Research (NIPER) – Guwahati, Assam, India

Shanta Biswas
Department of Applied Chemistry and Chemical Engineering,
Faculty of Engineering and Technology, University of Dhaka, Dhaka 1000, Bangladesh

Sunil Kumar Dubey
Department of Pharmacy, Birla Institute of Technology & Science (BITS), Pilani,
Rajasthan – 333031, India

Papia Haque
Department of Applied Chemistry and Chemical Engineering,
Faculty of Engineering and Technology, University of Dhaka, Dhaka 1000, Bangladesh

Md Saquib Hasnain
Department of Pharmacy, Shri Venkateshwara University, NH-24, Rajabpur, Gajraula,
Amroha, 244236, U.P., India

Santanu Kaity
Cadila Healthcare Limited, Ahmadabad, Gujarat, India

Moumita Ghosh Laskar
Department of Medicine, Karolinska University Hospital, Huddinge, Karolinska Institute,
Stockholm, Sweden

Abul K. Mallik
Department of Applied Chemistry and Chemical Engineering,
Faculty of Engineering and Technology, University of Dhaka, Dhaka 1000, Bangladesh

Shakti Nagpal
Department of Pharmacy, Birla Institute of Technology & Science (BITS), Pilani,
Rajasthan – 333031, India

Amit Kumar Nayak
Department of Pharmaceutics, Seemanta Institute of Pharmaceutical Sciences,
Mayurbhanj–757086, Odisha, India

Dilipkumar Pal
Department of Pharmaceutical Sciences, Guru Ghasidas Vishwavidyalaya
(A Central University), Koni, Bilaspur, C.G., 495009, India

A. Papagiannopoulos
Theoretical and Physical Chemistry Institute, National Hellenic Research Foundation,
48 Vassileos Constantinou Avenue, 11635 Athens, Greece

S. Pispas
Theoretical and Physical Chemistry Institute, National Hellenic Research Foundation,
48 Vassileos Constantinou Avenue, 11635 Athens, Greece

Mohammed Mizanur Rahman
Department of Applied Chemistry and Chemical Engineering,
Faculty of Engineering and Technology, University of Dhaka, Dhaka 1000, Bangladesh

Vamshi Krishna Rapalli
Department of Pharmacy, Birla Institute of Technology & Science (BITS), Pilani,
Rajasthan – 333031, India

Fiona Concy Rodrigues
Department of Biomedical Engineering, Manipal Institute of Technology,
Manipal Academy of Higher Education, Manipal–576104, Karnataka, India

Supriyo Saha
Department of Pharmaceutical Sciences, Sardar Bhagwan Singh University, Balawala,
Dehradun – 248001, Uttarakhand, India

Md Nurus Sakib
Department of Applied Chemistry and Chemical Engineering,
Faculty of Engineering and Technology, University of Dhaka, Dhaka 1000, Bangladesh

Md Shahruzzaman
Department of Applied Chemistry and Chemical Engineering,
Faculty of Engineering and Technology, University of Dhaka, Dhaka 1000, Bangladesh

Krizma Singh
Department of Biomedical Engineering, Manipal Institute of Technology,
Manipal Academy of Higher Education, Manipal – 576104, Karnataka, India

Gautam Singhvi
Department of Pharmacy, Birla Institute of Technology & Science (BITS), Pilani,
Rajasthan – 333031, India

Mohammad Tabish
Department of Pharmacology, College of Medicine, Saqra University – 1196,
Kingdom of Saudi Arabia

Goutam Thakur
Department of Biomedical Engineering, Manipal Institute of Technology,
Manipal Academy of Higher Education, Manipal – 576104, Karnataka, India

Abbreviations

°C	degree Celsius
APIs	active pharmaceutical ingredients
APTI	Association of Pharmaceutical Teachers of India
AZT	3'-azido-3'-deoxythymidine
BSA	bovine serum albumin
CAS	chemical abstracts service
CFR	code of federal regulations
CHC-GG	carboxymethyl chitosan
chitosan-TBA	chitosan-4-thio-butyl-amidine
cP	centipoise
CS	chitosan
Da	Dalton
DEE	drug encapsulation efficiency
DOT	dissolved oxygen and oxygen transfer capacity
DS	degree of substitution
EDC	1-ethyl-3-(3-dimethylaminopropyl) carbodiimide
g/mol	gram/mole
GA	glutaraldehyde
GAGs	glycosaminoglycans
GBR	guided bone regeneration
GG	guar gum
GIT	gastrointestinal tract
GTMAC	glycidyl trimethyl ammonium chloride
HAGG	high acyl gellan gum
HCl	hydrochloride
HPMC	hydroxypropyl methylcellulose
HTCC	N-(2-hydroxyl) propyl-3-trimethyl ammonium chitosan chloride
IND	indomethacin
IPN	interpenetrating polymer network
IUPAC	International Union of Pure and Applied Chemistry
J/g	joule per gram
LAGG	low acyl gellan gum

mPas	mili-Pascals
NNRTI	non-nucleoside reverse transcriptase inhibitor
PEI	polyethyleneimine
PLGA	poly(L-lactide-co-glycolide)
PM	paracetamol
PVA	polyvinyl alcohol
PVP	polyvinyl pyrrolidone
SCMC	sodium carboxymethyl cellulose
SGF	simulated gastric fluid
SIF	simulated intestinal fluids
TGF- $\beta1$	transforming growth factor $\beta1$
TMC	N-trimethyl chitosan
TPGS	tocopherol polyethylene glycol succinate 1000
USDA	United States Department of Agriculture
USFDA	United States Food and Drug Administration
XG	xanthan gum
β	beta
μm	micron

Preface

The polymers derived from various marine sources and microorganisms offer great industrial significance by contributing a decrease of the extraction costs. Most of these polymers possess some important biological properties such as biocompatibility, biodegradability, and bioadhesivity. Moreover, these polymers can be modified physically and/or chemically to improve their biomaterial properties. During past few decades, a number of marine- and microbiologically derived polymers have been explored and exploited as pharmaceutical excipients in various pharmaceutical dosage forms like tablets, microparticles, nanoparticles, ophthalmic preparations, gels, emulsions, suspensions, etc. The commonly used marine- and microbiologically derived polymers (used as pharmaceutical excipients) are alginates, agar-agar, gellan gum, carrageenan; chitosan (CS), xanthan gum (XG), etc.

This current volume of the book, *Natural Polymers for Pharmaceutical Applications, Volume 2: Marine- and Microbiologically Derived Polymers*, contains seven important chapters that present the latest research updates on the plant-derived polymers for various pharmaceutical applications. The topics of the chapters of the current volume include marine-derived polysaccharides: pharmaceutical applications of alginates, agar-agar, gellan gum, carrageenan, CS derivatives, and XG. This book particularly discusses the aforementioned topics along with an emphasis on recent advances in the field by the experts across the world.

We would like to thank all the authors of the chapters for providing timely and excellent contributions, the publisher Apple Academic Press (USA), and Sandra Sickels for the invaluable help in the organization of the editing process. We gratefully acknowledge the permissions to reproduce copyright materials from a number of sources. Finally, we would like to thank our family members, all respected teachers, friends, colleagues, and dear students for their continuous encouragements, inspirations, and moral support during the preparation of the book. Together with our contributing authors and the publishers, we will be extremely pleased if our efforts fulfill the needs of academicians, researchers, students, polymer engineers, and pharmaceutical formulators.

—Amit Kumar Nayak, PhD
Md Saquib Hasnain, PhD
Dilipkumar Pal, PhD

CHAPTER 1

Marine-Derived Polysaccharides: Pharmaceutical Applications

DILIPKUMAR PAL,[1] SUPRIYO SAHA,[2] AMIT KUMAR NAYAK,[3] and MD SAQUIB HASNAIN[4]

[1]*Department of Pharmaceutical Sciences, Guru Ghasidas Vishwavidyalaya (A Central University), Bilaspur, C.G., 495 009, India*

[2]*Department of Pharmaceutical Sciences, Sardar Bhagwan Singh University, Balawala, Dehradun – 248001, Uttarakhand, India*

[3]*Department of Pharmaceutics, Seemanta Institute of Pharmaceutical Sciences, Mayurbhanj, 757086, Odisha, India*

[4]*Department of Pharmacy, Shri Venkateshwara University, NH-24, Rajabpur, Gajraula, Amroha, 244236, U.P., India*

ABSTRACT

Alginate, carrageenan, and chitosan (CS) are the principle marine polysaccharides obtained from algae, sea sponge, sea horse, seaweed, and shrimps, etc. and observed with pharmaceutical applications. Alginate is used as a carrier molecule for controlled release of anticancer drugs, antiviral agents and also used as a wound healer. Carrageenan is used in controlled release carrier for curcumin, methotrexate, and also applied as intestine directed, vaginal insert and oral insulin delivering agent. CS is a marine polysaccharide, used as controlled release carrier of sunitinib, 5-fluorouracil, coumarin, and also observed with wound healing, trypanocidal, and mosquito larvacidal effects. This chapter provides detailed information about the pharmaceutical application of different marine polysaccharides.

1.1 INTRODUCTION

Marine polysaccharides are the natural polysaccharide derived from marine sources, such as algae, sponge, fish, seaweed, etc. (Rasummnen et al., 2007; Montaser et al., 2011). The principle importance of marine polysaccharides is that they are biologically compatible with low extraction cost. The characteristics of polymer were regulated by salt effect, temperature, pressure, presence of microbes, and viruses (Ahmad et al., 2014; Kim et al., 2008). Alginate, carrageenan, and chitosan (CS) are the main types of polysaccharides obtained from marine sources. Alginate is the sodium or calcium salt of alginic acid, an anionic polysaccharide obtained from *Laminaria hyperborea, Laminaria digitata, Laminaria japonica, Ascophyllum nodosum,* and *Macrocystis pyrifera.* Alginate is a polysaccharide developed upon the reaction of mannuronic acid and guluronic acid with biodegradability for promoted cell survival rate. The solubility and water retention capacity of alginate depends upon pH, molecular mass, and ionic strength of the polymer (Lee et al., 2012; Sachan et al., 2009). Carrageenan is a marine polysaccharide extracted from *Rhodophyceae* seaweed family having sulfur-containing polygalactan and has several types such as kappa, lambda, iota, epsilon, etc. as per its solubility in potassium chloride solution. Though lambda carrageenan is not capable of forming a gel, but iota carrageenan is able to form a right-handed helix. Carrageenan is mainly used as inflammation initiator in rat and responsible for anticoagulant, antithrombotic, antiviral antitumor, and other immunomodulatory activities (Necas et al., 2013; Tobackman et al., 2001). CS is a heteropolymer composed of glucosamine and *n*-acetyl glucosamine (Randy et al., 2015; Shweta et al., 2015; Nayak et al., 2011, 2012a). CS is extracted from chitin and mainly obtained from shells of lobster, shrimp, and crabs and treated with alkali or reacted with papain, CS is extracted (Nayak et al., 2012b, 2012c, 2014a, 2014b, 2014c). Hydrolytic cleavage of acetamide group, chitin is converted into CS with a greater degree of deacetylation (Nayak et al., 2015a, 2015b, 2016). In this chapter, we mainly emphasized on the pharmaceutical application of marine polysaccharides.

1.2 PHARMACEUTICAL APPLICATION OF ALGINATE

1.2.1 pH-RESPONSIVE ALGINATE HYDROGEL FOR DELIVERY OF PROTEIN MOLECULE

Lima et al., developed alginate-based hydrogel for the delivery of protein molecule (bovine serum albumin (BSA)). The formulation was developed by reaction of sodium alginate, glycidyl methacrylate, N-vinyl pyrrolidone, and sodium acrylate using tetramethylethylenediamine as catalyst (Table 1.1). FTIR, SEM, and NMR techniques were used to chemically characterize the hydrogel. Cytotoxicty evaluation against epithelial colorectal adenocarcinoma cells (HT-29) and swelling characteristic by gel permeation technique were evaluated for the formulation. SEM data revealed that at pH 7.4 formulations were more porous than pH 1.2. Cell viability data greater than 70% correlated with no inhibition of cell growth. At pH 1.2, formulation frequently achieved the pseudo-equilibrium state with loss of water molecules, and a pseudo Fickian distribution was observed with one of the formulations (Figure 1.1). This data clearly stated the importance of pH on the activity profile of the hydrogel for the delivery of BSA (Lim et al., 2018).

TABLE 1.1   Contents of Alg-GMA, VP, and SA Used in the Hydrogel-Forming Suspensions

Samples	Alg-GMA (g)	VP (mL)	SA (g)
Alg1V1A1	0.25	0.5	0.5
Alg1V1A2	0.25	0.5	1.0
Alg1V2A1	0.25	1.0	0.5
Alg1V2A2	0.25	1.0	1.0
Alg2V1A1	0.50	0.5	0.5
Alg2V1A2	0.50	0.5	1.0
Alg2V2A1	0.50	1.0	0.5
Alg2V2A2	0.50	1.0	1.0

Source: Lima et al., Copyright © 2018 with permission from Elsevier B.V.

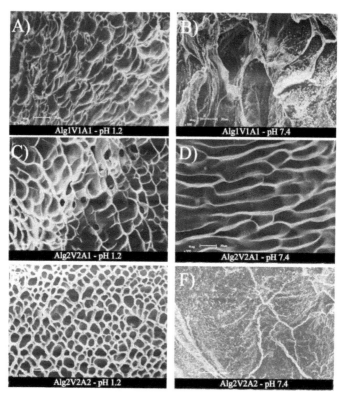

FIGURE 1.1 Micrographs of hydrogels with different amounts of GMA-Alg, *N*-vinyl-pyrrolidone, and sodium acrylate taken from samples freeze-dried after being swollen to equilibrium in water.

Source: Lima et al., Copyright © 2018 with permission from Elsevier B.V.

1.2.2 PLATINUM-ALGINATE NANOPARTICLES FOR THE TREATMENT OF HEPATIC CARCINOMA

Wang et al., developed a platinum (Pt IV) conjugated alginate nanoparticles for the treatment of hepatic carcinoma. The formulation was formulated by the reaction of alginate-tetrabutylammonium hydroxide, diammonium dihydroxy platinum (IV) dichloride, 2-Chloro-1-methylpyridinium iodide, ammonium glycyrrhetinic acid under nitrogen gas atmosphere at 0°C temperatures. FTIR, TEM, dynamic light scattering, and zeta potential were used to evaluate the chemical nature of the formulation. Drug loading efficiency, amount of glycyrrhetinic acid in formulation, MTT assay against HepG2 cell line and *in vitro* drug release were used as a pharmacological parameter to evaluate formulation. The characterization data obtained from TEM showed 141.9 nm of mean hydrodynamic diameter with spherical structure and -38.4 mV of zeta potential and total 120 h was needed to complete release of drugs. The MTT assay showed 8.73 µg/ml (IC_{50}). These data clearly confirmed the effectivity of nanoparticle in hepatic carcinoma (Wang et al., 2019).

1.2.3 SILICA-ALGINATE AEROGEL BEAD AS CONTROLLED RELEASE FORMULATION

Bugnone et al., developed an aerogel bead composed of silica and alginate by internal emulsion forming techniques. The formulation was developed by using the reaction of tetramethyl orthosilicate and sodium alginate solution, and the mixture was poured into the oil phase (paraffin and span 40 mixtures) in the presence of glacial acetic acid. After proper propelling and addition of deionized water, the hydrogel was developed, which was loaded with ketoprofen. Calcium-alginate or silica along with hydroxypropyl methylcellulose (HPMC) was used to coat the aerogel. Particle size distribution, drug loading, and *in vitro* drug release were the characterization parameters of formulations. The outcomes revealed that viscosity was gradually increase with tetramethyl orthosilicate and it was also observed that uncoated aerogel showed 95% release efficiency after 10 min, calcium alginate-hydroxypropylmethylcellulose with 80% efficiency in 35 min, silica-hydroxypropylmethylcellulose with 48% in 15 min; these data not satisfied with the release of ketoprofen release. Whereas in case of aerogel with hydrophobic silica coating, 60% release was observed after 30 min with good release kinetics of ketoprofen; which resembled this formulation with hydrophobic silica coating would become a good pharmaceutical carrier (Bugnone et al., 2018).

1.2.4 ALGINATE-HYALURONAN SPONGE AS DRESSING MATERIAL FOR POSTEXTRACTIVE ALVEOLAR WOUNDS

Catanzano et al., developed alginate-hyaluronan composite for the delivery of tranexamic acid for the dressing of post-extractive alveolar wounds. The formulation was developed by the reaction of macroporous alginate and hyaluronan (10% and 20%) embedded with tranexamic acid. The formulation was characterized by morphology, porosity, *in vitro* drug release and *in vitro* dynamic whole blood clotting parameters (Figure 1.2). The outcomes revealed that (300–500) μm particle size diameter with interconnected pores, fastest drug release, and blood-clotting index was 40% as compared to negative control. These data clearly stated the importance of alginate-hyaluronan sponge on the post-extractive alveolar wounds (Catanzano et al., 2017).

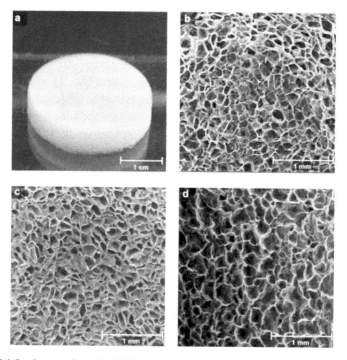

FIGURE 1.2   Image of an ALG/HA20 sponge after freeze-drying (scale bar ¼ 1 cm) (a). SEM images of ALG (b), ALG/HA10 (c), and ALG/HA20 (d) (scale bar ¼ 1 mm). *Source:* Catanzano et al., Copyright © 2018 with permission from Elsevier B.V.

1.2.5   IBUPROFEN LOADED SERICIN-ALGINATE FORMULATION AS SUSTAINED DRUG DELIVERY SYSTEM

Freitas et al., developed a mucoadhesive formulation made up of sericin and alginate for the sustained release of ibuprofen. Sericin was extracted from *Bombyx mori,* which was mixed with sodium alginate embedded with ibuprofen. Drug incorporation efficiency and *in vitro* drug release were used to characterize the formulation. The drug incorporation efficiency was (73.01–94.15) and % drug release was regulated by alginate concentration with Korsmeyer-Peppas kinetic model and (0.80 ± 0.13 to 1.08 ± 0.11) mm of particle size (Figure 1.3). These data clearly stated the importance of ibuprofen-loaded formulation for the sustained delivery of a drug molecule (Freitas et al., 2018).

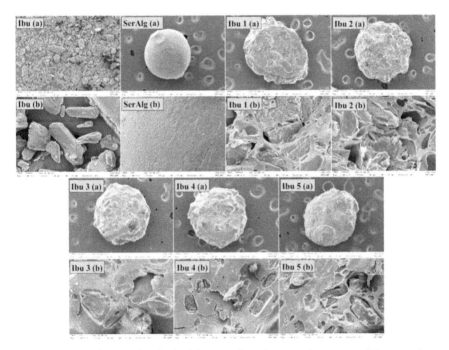

FIGURE 1.3 Micrographs obtained by scanning electron microscopy of the pure drug (Ibu = ibuprofen), sericin/alginate particles (SerAlg), and sericin/alginate particles loaded with ibuprofen (Ibu 1–Ibu 5) at magnifications of (a) 150× and (b) 1000×.

Source: Freites et al., Copyright © 2018 with permission from Elsevier B.V.

1.2.6 ALGINATE HYDROGEL FOR CONTROLLED RELEASE OF VANILLIN

Garcia-Astrain et al., developed pH sensitive alginate hydrogel for the controlled release of vanillin with palpitate swelling behavior. The formulation was developed upon the reaction of the carboxylic acid group of alginate and furfurylamine, which was reacted with bismaleimide to obtain hydrogel. FTIR, [1H]NMR, percent swelling ration, and percent cumulative release of vanillin were used to characterize the formulation. Hydrogel with the highest degree of bismaeimide substitution was observed with an interconnected porous structure with 230 μm pores and higher viscosity. Maximum swelling ratio of hydrogel with a higher concentration of bismaleimide was observed at pH 1.3, and it was correlated that swelling ratio was gradually increased with the deprotonated carboxylate group of alginate; even after 3–4 h maximum drug was released from formulation. This data stated the importance of this alginate hydrogel for the controlled release of vanillin (Garcia-Astrain et al., 2018).

1.2.7 ALGINATE HYDROGEL AS CONTROLLED RELEASE CARRIER OF IMATINIB MESYLATE

Jalababu et al., developed dual responsive alginate-poly acrylamide hydrogel for controlled release of imatinib mesylate. The formulation was developed upon polymerization of NaAlg-g-PAPA copolymer (a polymer of sodium-alginate and acryloyl phenylalanine), N-isopropyl acrylamide and N,N-bismethylenebisacrylamide in the presence of ammonium persulphate. The formulation was chemically characterized by FTIR, SEM, DSC, and XRD; pharmaceutical characterization was done by swelling parameters, *in vitro* drug release. The outcomes revealed that 79% of encapsulation and 98% of release were observed after 48h (pH 7.4, 37°C) with the Korsmeyer-Peppas kinetic model. These data clearly informed the controlled release characteristic of alginate hydrogel using imatinib mesylate as a model drug (Jalababu et al., 2018).

1.2.8 ZIDOVUDINE EMBEDDED ALGINATE FORMULATION FOR ANTI-HIV DRUG DELIVERY

Joshy et al., developed pluronic F68 coated alginate nanoparticle embedded with zidovudine with the effectivity against human immunodeficiency virus. The formulation was developed using the reaction of zidovudine and glutamic acid- alginate conjugate in dichloromethane; whereas the conjugate was coupled using sodium alginate, glutamic acid, 1-ethyl-3-(3-dimethylaminopropyl) carbodiimide (EDC) hydrochloride (HCl) and *n*-hydroxysuccinimide at room temperature. The primary emulsion was poured into a water phase containing PluronicF-68 as a surfactant, further centrifuged at 20,000 rpm for 30 min. The formulation was characterized by FTIR, TEM, Polyacrylamide gel electrophoresis, *in vitro* drug release and MTT assay against C6 glioma and neuro 2a cell lines. The outcomes showed TEM data with 432 nm of particle size, 74.2% of encapsulation efficiency, and here within 24 h, 54% drug was released from nanoparticle (Figure 1.4 and Table 1.2). MTT assay observed with good cytotoxicity. This information indicated zidovudine embedded with alginate nanoformulation for the treatment of HIV (Joshy et al., 2017).

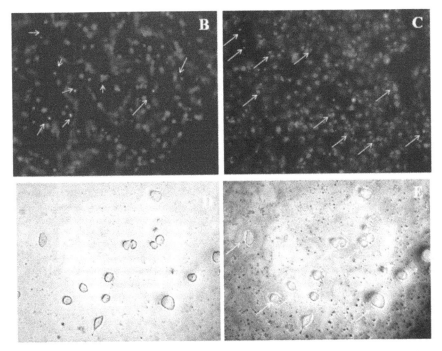

FIGURE 1.4   (See color insert.) Fluorescent microscopic examination of C6 cell lines and neuro 2a cell lines treated with (B) RhB labeled AZT-GAAD NPs (C) merged image of AZT-GAAD NPS with cell nuclei stained blue with Hoechst 33342(D) bright field image of neuro 2a cells(E) RhB labeled AZT-GAAD NPs.

Source: Joshy et al., Copyright © 2018 with permission from Elsevier B.V.

TABLE 1.2　Particle Size, PDI, and Zeta Potential of AZT-GAAD Nanoparticles

Formulation code	Particle size (nm)	PDI	Zeta potential (mV)
AZT-GAAD NPs1	273±2.6	0.2±0.004	−42.16±3.2
AZT-GAAD NPs2	267±2.3	0.3±0.006	−36.42±2.7
AZT-GAAD NPs3	237±2.3	0.3±0.002	−34.13±1.61

Source: Joshy et al., Copyright © 2018 with permission from Elsevier B.V.

1.2.9　CIPROFLOXACIN LOADED ALGINATE FIBROUS FORMULATION FOR WOUND HEALING

Liu et al., developed an electrospun fibrous mat composed of sodium alginate-polylactic-co-glycolic acid embedded with ciprofloxacin as wound dressing material. The formulation was developed using the reaction of polylactic glycolic acid in electrospun in (4:1) v/v of trifluoroethanol and chloroform mixture. Ciprofloxacin was embedded into the fibrous mats and the mixture of sodium alginate to get the final formulation. The formulation was characterized by morphology assessment by SEM, drug release behavior, biodegradation behavior, and antimicrobial activity against *S. aureus*. The outcomes revealed that particle size distribution was 100 nm to 15 μm, the water contact angle was lower than 66°. At pH 7.4 in phosphate buffer solution, burst release was observed within 7 days. A slow release of ciprofloxacin was noticed for 40–50 days, and electrospun formulation with 4% sodium alginate showed best antimicrobial efficacy. This formulation was observed with greater efficiency for the dressing of wound healing (Liu et al., 2018).

1.2.10　MINOXIDIL INCLUSION COMPLEXED ALGINATE HYDROGEL FOR ALOPECIA TREATMENT

Lopedota et al., developed an alginate hydrogel embedded with an inclusion complex of beta-hydroxy cyclodextrin and minoxidil for the effective treatment of alopecia. The formulation was developed upon the reaction of minoxidil 1–6% and 0.65 beta-hydroxy cyclodextrins along with 0.5 g sodium alginate and poured the mixture into calcium chloride solution. The formulation was characterized by the release of minoxidil from hydrogel, *ex vivo* skin study, histological study as major parameters. The outcomes revealed that after 24 h, 70.80% of minoxidil was released, 65.5 µg/cm^2 of best accumulation was observed by *ex vivo* skin accumulation test as compared to minoxidil marketed solution and a widespread cell hyperplasia with altered desquamation was observed by histological studies. These data clearly suggested the importance of the formulation for alopecia treatment (Lopedota et al., 2018).

1.2.11 SPRAY DRIED ALGINATE AS CARRIER FOR ROFLUMILAST DELIVERY

Mahmoud et al., developed alginate carrier by emulsified spray dried technique for the delivery of roflumilast active against A549 lung cancer cell line. The formulation was developed upon mixing of Tween 80 (0.05% w/v), sodium alginate (0.25% w/v), aqueous solution (0.5% w/v) of lactose, mannitol, beta-cyclodextrin, and maltodextrin along with roflumilast (0.05% w/v). In the final step using shock heating-cooling phenomenon with standard temperature and pressure nanoparticles were formed. The morphology of the formulation was characterized by SEM, XRD, DSC, particle size diameter, and in vitro drug release. MTT assay against A549 was the main characterization parameter. The outcomes revealed that 79.5% encapsulation efficiency, (2.14–2.57) µm of particle size, 28.35%/h of the drug was released from the formulation with cyclodextrin with release up to 8 h and potent inhibition against A549 cell line. These data clearly stated that alginate formulation with beta-cyclodextrin as crosslinker would be the best formulation (Mahmoud et al., 2018).

1.2.12 QUERCETIN DELIVERY BY CHITOSAN (CS)-ALGINATE NANOPARTICLE FOR ANTI-DIABETES THERAPY

Mukhopadhyay et al., developed pH sensitive CS-alginate corona structured nanoparticle embedded with quercetin for the treatment of diabetes. The nanoformulation was developed upon reaction between succinic anhydride treated CS and alginate and poured into calcium chloride solution incorporated with quercetin solution (1 mg/ml). FTIR, 1HNMR, XRD, particle size determination, drug loading, and encapsulation efficiency along with antidiabetic activity, were the principal characterization parameter. The particle size of the nanoformulation was 91.58 nm. It possessed zeta potential of ((–)19.83–(–)35.55), 90% of encapsulation efficiency, and 59% of drug loading with a significant hypoglycemic response. This statement indicated the polymeric delivery of quercetin for the treatment of hyperglycemia (Mukhopadhyay et al., 2018).

1.2.13 PROTEIN SANDWICHES ALGINATE COMPOSITE FOR ION SENSITIVE DELIVERY OF PROTEIN

Rahmani et al., developed a composite of protein (cytochrome C, lysozyme, myoglobin, chymotrypsin, and BSA) and alginate as the carrier of the protein molecule. The formulation was developed using different protein and alginate in the sodium acetate buffer solution at pH 4.5. Bradford assay to characterize the protein measurement and release were measured at pH 4.5, 7.4 in sodium acetate and phosphate buffer, respectively, with kinetic assessment (Table 1.3). Lysozymes and cytochrome C were the most complexing agent with alginate, and ionic strength & pH of the medium were the regulating parameters for the release of protein. These data clearly indicated the importance of cytochrome and lysozyme interlocked with alginate would help to deliver the protein molecule (Rahmani et al., 2018).

TABLE 1.3 Isoelectric Point, Net Charge at Physiological pH, and Molecular Weight of the Model Proteins

Protein	pI / Net Charge	Molecular Weight (Da)
Cytochrome C	10.0 / Positive	12000
Lysozyme	11.0 / Positive	14000
Myoglobin	7.2 / Neutral	17000
Chymotrypsin	9.1 / Positive	25000
Bovine Serum Albumin (BSA)	5.4 / Negative	65000

Source: Rahmani et al., Copyright © 2017 with permission from Elsevier B.V.

1.2.14 LIPID-ALGINATE NANOPARTICLE FOR DELIVERY OF AMPHOTERICIN B

Senna et al., developed lipid and alginate nanostructure for the delivery of amphotericin B. The nanoformulation was developed using homogenizer upon reaction of glyceryl monostearate, mygliol 812N, span 80 and amphotericin B along with Tween 80 and kolliphor P188 and the formulation was merged with different concentration of alginate (1.5%, 2.0%, 2.5%), followed by poured in calcium chloride solution. Encapsulation efficiency, swelling studies, and cytotoxic effect against Vero cell were the principal characterization parameters. The particle size of the formulation was 96.97 µm without any cell toxic effects with the satisfied swell property. These data clearly stated the lipid-alginate nanocomposite for the delivery of amphotericin B (Senna et al., 2018).

1.2.15 SUBSTITUTED ACRYLAMIDE GRAFTED ALGINATE AEROGEL AS CONTROLLED RELEASE CARRIER

Shao et al., developed an aerogel composed of N-isopropylacrylamide and N-hydroxymethylacrylamide grafting with sodium alginate, ammonium persulphate as reaction initiator and N,N,N,'N'-tetramethylethylene-diamine as reaction accelerator, which was finally mixed with gluconic acid δ-lactone to obtain aerogel. Polydispersity index, drug loading, and release were the principle parameter to characterize the aerogel. It was found that the polydispersity index was 1.56 with release behavior maintained the Higuchi and first-order model with the acceleration in drug release observed at 42°C (Figure 1.5). These data indicated the importance of smart aerogel as a controlled release carrier (Shao et al., 2018).

1.2.16 CURCUMIN DIGLUTARIC ACID DELIVERY SYSTEM WITH CHITOSAN (CS)-ALGINATE FOR ANTICANCER THERAPY

Sorasitthiyanukarn et al., used an old concept of CS coated alginate nanoparticle for the delivery of curcumin diglutaric acid (a prodrug of curcumin) for effective anticancer therapy. CS coated alginate nanoparticle was developed by ionotropic gelation method of CS, sodium alginate, Pluronic F-127 and curcumin diglutaric acid, which was characterized by drug encapsulation-loading efficiency, *in vitro* drug release and anticancer efficacy against Human epithelial colorectal adenocarcinoma, human hepatocellular carcinoma and human breast cancer cell lines. The outcomes revealed that the concentration of Pluronic F-127 was inversely proportional with encapsulation efficiency with greater effects against cell lines, especially epithelial colorectal adenocarcinoma cell line. This data stated that the curcumin diglutaric acid molded within nanocomposite would work as anticancer therapy (Sorasitthiyanukarn et al., 2018).

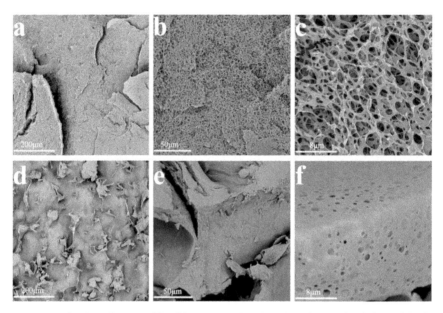

FIGURE 1.5   SEM images of intelligent aerogel (a, b, and c); indomethacin loaded intelligent aerogel.
Source: Shao et al., Copyright © 2018 with permission from Elsevier B.V.

1.2.17   HYDROXYAPATITE NANOCOMPLEXED ALGINATE AS CONTROLLED RELEASE CARRIER

Sukhodub et al., developed nanocomposite composed of hydroxyapatite and sodium alginate for the controlled release behavior of chlorhexidine as an antimicrobial agent. The formulation was developed upon chemical reaction followed by centrifugation of alginate, hydroxyapatite, calcium nitrate tetrahydrate, and in the final formulation, pH was adjusted to 10.5. Here three types of samples were prepared using drying, lyophilization, and annealing techniques with composite formation using chlorhexidine. XRD, SEM, TEM, release kinetics were the principle parameters for characterization. The observation indicated that at 80 nm particle size, the higher crystalline arrangement was noticed and release kinetics revealed that lyophilized formulation showed better release than others at 37°C in phosphate buffer solution. These data indicated the importance of this nanocomplex for the controlled release of chlorhexidine molecule (Sukhodub et al., 2018).

1.2.18   SODIUM ALGINATE-POVIDONE IODINE COMPOSITE FILM FOR CONTROLLED ANTISEPTIC RELEASE

Summa et al., developed a composite film of sodium alginate and povidone iodine for the controlled release antiseptic treatment. The film was developed using sodium alginate and povidone iodine with constant stirring at room temperature for 1h. The film of 55 cm^2/ml was used to characterize the film for cell toxicity and anti-inflammation characteristics using HFF-1 (Human Frozen Fibroblast) cells. In-vivo wound healing and cytokine expression measurements were also evaluated using mice. Lower inflammation with greater survival rates and greater wound healing behavior were the primary outcomes; whereas progressive hydroxyproline concentration indicates slowly rebuild of the epithelium. These data stated the importance of this alginate-povidone iodine film for antiseptic treatment (Summa et al., 2018).

1.2.19   NANOCRYSTAL HYBRID OF ALGINATE AND CELLULOSE FOR CONTROLLED RELEASE OF RIFAMPICIN

Thomas et al., developed a nanocrystalline hybrid of alginate and cellulose (extracted from banana fiber) using honey as stabilizer followed by ionotropic gelation for the oral controlled release behavior of rifampicin in the inhibition of *Mycobacterium tuberculosis*. The formulation was developed by sonication of sodium alginate, cellulose nanocrystals and rifampicin at 3500 rpm for 5 mins. Entrapment efficiency, *in vitro* drug release and cytotoxicity against the L929 cell line, were the main characterization parameters. The outcomes came with (43.82–69.73)% of entrapment efficiency, (70–100) nm of particle size along with ((–)15–(–)20 mV) as zeta potential for all the formulations (Figure 1.6). The *in vitro* drug release data revealed that at pH 7.4, 100% rifampicin was released with approximate 100% cell viability index. These data clearly indicated that this nanocrystalline hybrid would become major controlled release carrier of rifampicin (Thomas et al., 2018).

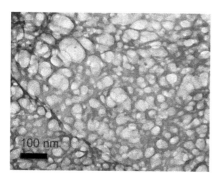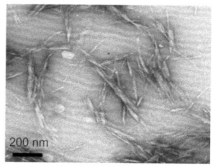

FIGURE 1.6   A) TEM of the ALG-CNC NPs and (b) TEM of the CNCs.

Source: Thomas et al., Copyright © 2018 with permission from Elsevier B.V.

1.2.20   ESSENTIAL OIL ENCAPSULATED ALGINATE-SOY PROTEIN SYSTEM FOR TARGETED INTESTINAL DELIVERY

Volic et al., developed an intestinal delivery system of thyme oil composed of alginate and soy protein. The formulation was developed using the reaction between emulsion of denatured soy protein and sodium alginate with thyme oil (1:5) ratio, which was poured into calcium chloride solution for bead formation. The formulation was mainly characterized by encapsulation efficiency, release studies in gastrointestinal mode, swelling properties as the main parameters. The outcomes revealed that (72–80)% of encapsulation efficiency with (42–55)% of drug release. These data clearly stated that the alginate-soy protein system would suitable for intestinal delivery of thyme oil (Volic et al., 2018).

1.2.21   AGAR-ALGINATE HYDROGEL AS ORAL DRUG DELIVERY CARRIER SYSTEM

Yin et al., developed a pH-responsive hydrogel system composed of agar and alginate for the oral delivery of drug molecules. The formulation was developed upon the reaction of sodium alginate, agar, indomethacin, and calcium chloride as a bead-forming agent. Swelling characteristic at suitable pH and *in vitro* drug release at pH 7.4, and 2.1 at 37°C were the main characterization parameters. This revealed that a greater concentration of agar was correlated with lower swelling and the complete drug was released within 12 h. These data stated the development of agar-alginate composite hydrogel for oral delivery (Yin et al., 2018).

1.2.22 ℰ-POLYLYSINE-ALGINATE COMPOSITE NANOPARTICLE FOR SUSTAINED DELIVERY OF ANTIGEN

Yuan et al., developed ℰ-polylysine-alginate nanocomposite particle embedded with BSA for the delivery of antigen. The formulation was developed upon the reaction of ℰ-polylysine, sodium alginate and BSA and filtered through 0.45µ filter paper. Encapsulation efficiency, *in vitro* drug release in phosphate buffer solution at pH 7.4 and *in vitro* cell toxicity against RAW264.7 cell line were the main characterization parameters. Outcomes revealed that 133.2 nm particle size without any cell toxicity and sustained release behavior was observed unto 5 days. It stated the benefits of sustained delivery of antigen composed with ℰ-polylysine and sodium alginate (Yuan et al., 2018).

1.3 PHARMACEUTICAL APPLICATION OF CARRAGEENAN

1.3.1 OVALBUMIN COMPLEXED κ-CARRAGEENAN NANOPARTICLE FOR THE DELIVERY OF CURCUMIN

Xie et al., developed a nanoparticle composed of ovalbumin and κ-Carrageenan for the delivery of curcumin. The formulation was developed upon the reaction of ovalbumin and κ-Carrageenan emulsion with an ethanolic solution of curcumin and finally centrifuged at 10,000 rpm for 20 min. The stability of the emulsion, determination of curcumin, particle size determination, and *in vitro* drug release were the main characterization parameters. Near isoelectric point (4.7) of ovalbumin, κ-carrageenan presence inhibited the flocculation of ovalbumin, reflected the importance of the carrier. The emulsifying stability was 84.6 min at 36 h, and after thermal effect at pH 6 and 7, system guarded the curcumin release; 70.1% of curcumin was released in *in-vitro* intestinal fluid followed by slow release. These outcomes confirmed that this formulation would better work for curcumin delivery (Xie et al., 2019).

1.3.2 EXTENDED RELEASE OF ZALTOPROFEN FROM STARCH-CARRAGEENAN HYDROGEL

Sonawane et al., developed a hydrogel composed of starch and κ-carrageenan for the extended release of zaltoprofen as an anti-inflammatory agent. The formulation was developed upon the reaction of starch and κ-carrageenan in (10% w/v glyoxal solution) and amalgamating with pellet forming agent (MCC PH 101); characterized by *ex vivo* mucoadhesive property, *in vitro*, and *in vivo* drug release parameters. The outcomes revealed that (91–98)% mucoadhesiveness was observed up to 12 h, *in vitro* drug release at pH 6.8. Pellet forming agent was not participated in sustained release behavior, and one of the formulations (4:2 ratio of starch and κ-carrageenan) was provided extended-release up to 12h (Figure 1.7). These outcomes revealed the importance of starch-κ-carrageenan composite hydrogel for the extended release of zaltoprofen (Sonawane et al., 2018).

1.3.3 CARRAGEENAN IRRADIATED WITH GAMMA RAYS AS POTENTIAL ANTIOXIDANT AGENT

Abad et al., developed a redefined κ-carrageenan polymer irradiated with gamma rays obtained from Co-60 (10 KGy/h) dose rate and (1)% of kappa, iota, and lambda. Carrageenan was also irradiated with gamma rays (5–50) KGy, followed by evaluation of antioxidative properties by hydroxyl radical scavenging, free radical scavenging and reducing power assay methods. The outcomes revealed that lambda carrageenan showed maximum antioxidant effect followed by iota carrageenan and kappa carrageenan; points of sulfation showed positive effects on reducing the effect of carrageenan. This data informed the importance of gamma rays irradiated carrageenan in radical scavenging effects (Abad et al., 2013).

1.3.4 PREVENTIVE ACTION AGAINST ENTEROVIRUS 71 BY CARRAGEENAN

Chiu et al., suggested the appreciable prevention of growth against Enterovirus 71 by carrageenan. New strain of Enterovirus 71 and Vero Cells were

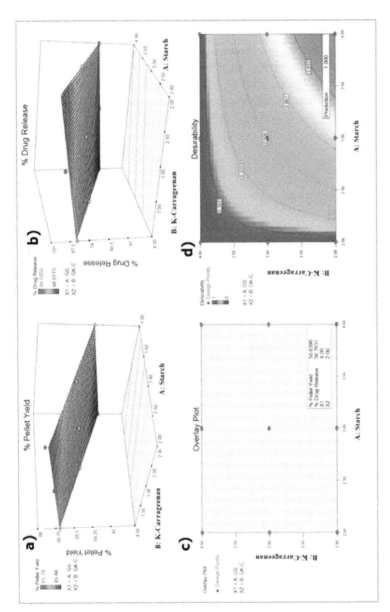

FIGURE 1.7 **(See color insert.)** Response surface plots showed the influence of starch concentration (X1) and κ- carrageenan concentration (X2) on (a) % Pellet size (Y1), (b) % Drug release (Y2), (c) overlay plot, and (d) desirability plot.

Source: Sonawane et al., Copyright © 2018 with permission from Elsevier B.V.

cultured in RPMI-1640 medium with fetal bovine serum (2%) at 37°C temperature followed by stored at (–)80°C temperature for evaluation by carrageenan polymer against cell proliferation assay by MTS, plaque reduction rate, binding of carrageenan with Enterovirus 71 as detected by ELISA (Figure 1.8). The outcomes revealed that carrageenan was observed with great effectiveness in plaque reduction and inhibition of enterovirus-induced apoptosis with the firm binding of carrageenan with a wall of enterovirus. These outcomes revealed the prevention of carrageenan against Enterovirus 71 with full zeal (Chiu et al., 2012).

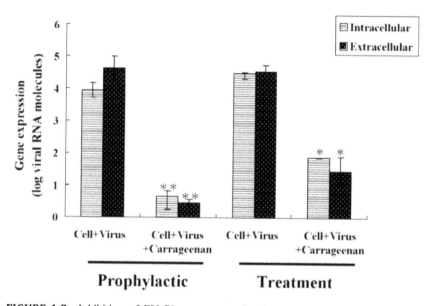

FIGURE 1.8   Inhibition of EV 71 gene expression by carrageenan. The prophylactic group: Vero cells were pre-treated with carrageenan and then infected with EV 71. The treatment group: Vero cells were pre-infected by EV 71 and then treated with carrageenan. *Source:* Chiu et al., Copyright © 2012 with permission from Elsevier B.V.

1.3.5 ENCAPSULATED ENTERIC DELIVERY OF PROBIOTIC BACTERIA WITH CARBOXYMETHYL CELLULOSE AND CARRAGEENAN COMPOSITE

Dafe et al., developed an enteric delivery of probiotic bacteria composed of carboxymethyl cellulose and carrageenan. The formulation was developed upon reaction of aqueous solutions of carboxymethyl cellulose and kappa-carrageenan and accelerated the temperature up to 70°C and poured into cross-linking solution (CaCl$_2$/KCl) and microencapsulated *L. plantarum* cells (10.01 Log CFU/g) into the bead, finally characterized by swelling studies, release statistics in simulated gastric fluid (SGF) and bile salts. The observations revealed that after exposing to pH 1.2, unbound cells, maximum abolished with 7.30 (Log CFU/g) were survived. This information stated the high effectiveness of carboxymethylcellulose-(k) carrageenan composite loaded with probiotic bacteria in a colon-targeted delivery system (Dafe et al., 2017).

1.3.6 IOTA CARRAGEENAN AND GELATIN ENCAPSULATE FOR TARGETED INTESTINAL DELIVERY

Gomez-Mascaraque et al., developed an encapsulated structure composed of ɩ carrageenan and gelatin with grape juice extract for the intestinal delivery. The formulation was blended upon the reaction of carrageenan and gelatin (with different ratio) embedded with extract and characterized by pH measurement, conductivity measurement, single angle X-ray scattering, the release of extract from the lyophilized formulation. At alkaline pH, maximum complexation, better entrapment efficiency, and lesser hygroscopicity were observed with (85:15) ratio of ɩ-carrageenan and gelatin. The statement indicated that the formulation would be effective for intestinal delivery of drug molecules (Gomez-Mascaraque et al., 2018).

1.3.7 GRIFFITHSIN COMPOSED CARRAGEENAN VAGINAL INSERT FOR SEXUALLY TRANSMITTED DISEASE TREATMENT

Lal et al., developed a fast dissolving vaginal insert by free drying technique of griffithsin (4 mg of loading dose) loaded carrageenan for the treatment of sexually transmitted disease and the formulation was characterized by content assessment of griffithsin and carrageenan, *in vitro* antiviral assay against Human Immunodeficiency Virus and Human Papilloma Virus, respectively. Outcomes showed that content of griffithsin were intact up to 6 months at 40°C and 75% with greater inhibition against immunodeficiency, papilloma, and herpes simplex virus. The statement concluded with greater efficiency against sexually transmitted disease (Lal et al., 2018).

1.3.8 ALGINATE AND K-CARRAGEENAN HYDROGEL FOR ORAL INSULIN DELIVERY

Lim et al., developed a hydrogel for controlled delivery of insulin as part composed with alginate and κ-carrageenan. The formulation was developed upon equal mixing of alginate (2% w/v)/κ-carrageenan (0–1)% w/v stock solution and insulin as a part solution (0.875 mg/ml) and dropwise poured into the calcium chloride solution (Figure 1.9). Encapsulation efficiency and *in vitro* release kinetics in simulated gastric and intestinal fluid of insulin as part were the principle characterization of the formulation. The outcomes revealed that encapsulation efficiency was (19–28)% and at pH 1.2, all insulin was retained, but in pH 7.4, insulin was stepwise released up to 2 h. So, the formulation would become a milestone in the delivery of insulin as part (Lim et al., 2017).

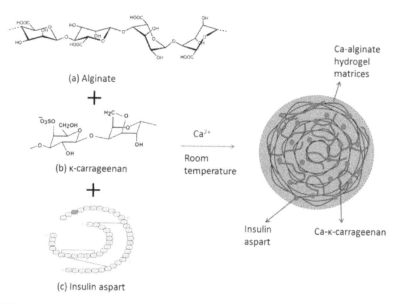

FIGURE 1.9 **(See color insert.)** Schematic diagram of the formation of insulin as part-loaded alginate/κ-carrageenan composite hydrogel beads.

Source: Lim et al., Copyright © 2017 with permission from Elsevier B.V.

1.3.9   κ-CARRAGEENAN AND CHITOSAN (CS) COMPLEX AS CONTROLLED RELEASE CARRIER OF METHOTREXATE

Mahdavinia et al., developed a pH sensitive and magnetic hydrogel of κ-carrageenan (prepared with $Fe^{2+}/Fe^{3+} = 2.0$) and CS for the release of methotrexate (20 mg in 20 ml in sodium hydroxide) and stirred at 800 rpm at 60°C for 20 min as controlled release carrier (Table 1.4). The formulation was characterized by encapsulation efficiency, swelling studies, and *in vitro* drug release parameters. Outcomes came with the observation that 67.5% and 63.5% of CS and magnetite were responsible for methotrexate release and approximate 68% of the drug was released at pH 7.4 with Fickian diffusion characteristics. The information stated with the importance of CS-carrageenan complex for the controlled release of methotrexate as anticancer delivery (Mahdavinia et al., 2017).

TABLE 1.4   Required Amount of Initial Materials to Prepare Magnetic κ-Carrageenan/Chitosan Hydrogels

	Chito0.4 Car	LmChito 0.4Car	HmChito 0.4Car	LmChito 0.1Car	LmChito 0.1Car	LmChito 0.6Car
κ-Carrageenan (g)	1	1	1	1	1	1
$FeCl_3.6H_2O$ (g)	0	1.35	2.7	1.35	1.35	1.35
$FeCl_2.4H_2O$ (g)	0	0.5	1	0.5	0.5	0.5
Chitosan (g)	0.4	0.4	0.4	0.1	0.2	0.6
κ-Carrageenan (g)	1	1	1	1	1	1

Source: Mahdavinia et al., Copyright © 2017 with permission from Elsevier B.V.

1.3.10   THIOLATED CARRAGEENAN AS CACO-2 CELL INHIBITOR

Suchaoin et al., developed a thiolated carrageenan (developed from the substitution of -OH group with -SH group, upon the reaction of brominated-carrageenan with thiourea followed by lyophilized at 4C temperature, -SH group was determined by Ellman's reagent) as an inhibitor of Caco-2 cell and efflux pump. Outcomes revealed that within (3–24) h, formulation showed concentration-dependent cell toxicity against colon cancer cell line, Sulforhodamine 101 dyed κ and ι carrageenan showed 1.38 and 1.35

greater than control. The development stated the importance of thiolated carrageenan as Caco-2 cell Inhibitor (Suchaoin et al., 2015).

1.4 PHARMACEUTICAL APPLICATION OF CHITOSAN (CS)

1.4.1 CROSSLINKED CHITOSAN (CS) AS CONTROLLED RELEASE CARRIER OF SUNITINIB

Karimi et al., developed a pH-responsive carrier composed of κ-carrageenan (0.2 g in 100 ml distilled water) and magnetically sound CS (1 g of CS dissolved in 100 ml glacial acetic acid and amalgamated with hydrated ferrous chloride and ferric chloride [$nFe^{2+}/nFe^{3+} = 2.0$) solu-tion], followed by centrifugation for the controlled release of sunitinib anticancer agent. The formulation was evaluated *in vitro* drug loading and *in vitro* drug release in phosphate buffer solution at pH 7.4, 4.5, 5.8 respectively, swelling studies and encapsulation efficiency. Outcomes revealed that all the formulations were observed with paramagnetic prop-erties. Swelling of the formulation was dependent upon the molecular weight of CS with (68.32–78.42)% of encapsulation efficiency, and drug release was favored in the acidic environment and protonation nature of CS. These data clearly stated the importance of this formulation to deliver sunitinib (Karimi et al., 2018).

1.4.2 CHITOSAN (CS)-BENTONITE NANOSTRUCTURE COMPOSITE AS WOUND HEALER

Devi et al., developed a nanocomposite film composed by mixing of CS solution (1% w/v in acetic acid) and bentonite (0.5%) and dried in a Petri dish at 60°C as potential wound healer, characterized by dressing material pH, antibacterial efficacy against *Bacillus subtilis* (Gram-positive) and *E. coli* (Gram-negative) strain, *in vitro* nonenzymatic hydrolytic degradation and hemocompatibility test (Scheme 1.1). Outcomes revealed that nano-composite film was provided with an acidic shield overwound to attenuate the cell growth. One formulation with 40 ml of CS and 10 ml of bentonite showed maximum antibacterial activity with greater hydrolytic degrada-tion. This information indicated the future of CS-bentonite nanofilm as great wound healer (Devi et al., 2017).

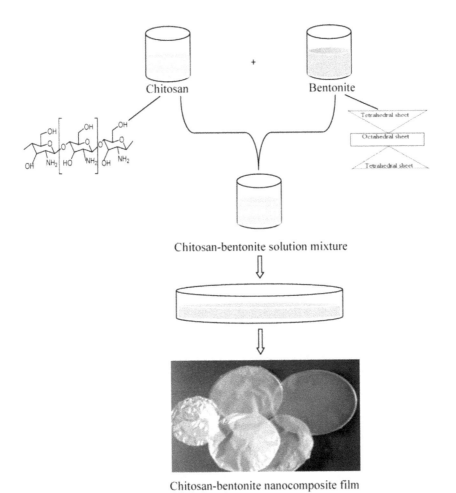

Chitosan-bentonite nanocomposite film

SCHEME 1.1 (**See color insert.**) A schematic representation for the preparation of chitosan-bentonite nanocomposite film.

Source: Devi et al., Copyright © 2017 with permission from Elsevier B.V.

1.4.3 CHITOSAN (CS) CONJUGATED FERULIC ACID AS POTENTIAL DRUG DELIVERY MATERIAL

Li et al., developed a conjugate of CS and ferulic acid under ascorbic acid and hydrogen peroxide in an inert redox pair environment for the delivery of drug material. The conjugate was developed upon the reaction of CS (10 g in 1% acetic acid), 20 ml 1M of hydrogen peroxide and 1.2 g of ascorbic acid and stirred constantly with ferulic acid (5 g), which was amalgamated with BSA as cell wall material. The grafting of the conjugate was evaluated by Folin-ciocalteu procedure, observed with (1.0:0.5) mass ratio of CS and ferulic acid. The encapsulation efficiency and *in vitro* drug release at sodium phosphate buffer (pH 7.4) were also evaluated; came with the observation that maximum albumin was released after 15h. This information stated the importance of the conjugate as potential drug delivery material (Li et al., 2017).

1.4.4   SUSTAINED RELEASE DRUG DELIVERY USING NITROSALICYLALDEHYDE AND CHITOSAN (CS) HYDROGEL

Craciun et al., developed a formulation composed of nitrosalicylaldehyde and CS using diclofenac sodium (model drug) as long term sustained release delivery system. The formulation was developed using the reaction between an ethanolic solution of nitrosalicylaldehyde and diclofenac sodium with an acetic acid solution of CS (2% w/v); characterized by *in vitro* drug release with kinetics, *in vitro* enzymatic degradation, *in vivo* drug release and biocompatibility. The outcomes revealed two formulations contained with (42.3 mg/0.213 mol), nitrosalicylaldehyde (18.2 mg/0.109 mol) and diclofenac sodium (1.5 mg) and CS (51.6 mg/0.260 mol), nitrosalicylaldehyde (8.9 mg/0.053 mol) and diclofenac sodium (1.5 mg) were most effective dosage forms having sustained release capacity within 8 days. This information stated the importance of composite hydrogel using nitrosalicylaldehyde and CS possessing long term sustained releasing behavior (Craciun et al., 2019).

1.4.5   CHITOSAN (CS) LINKED 5-FLUOROURACIL PRODRUG AS CONTROLLED DRUG DELIVERY CARRIER

Horo et al., developed a nanoparticle upon the reaction of low molecular weight CS (0.0019 mol) and bis (trimethylsilyl) acetamide linked 5-fluorouracil, followed by addition of sodium tripolyphosphate and lyophilized; further characterized by field emission transmission electron microscopy, dynamic light scattering and *in vitro* drug release parameters. Outcomes revealed that particle size with 70 nm to 90 nm, 53% of drug release after 4h, and 0.127 of polydispersity index were optimum for drug delivery carrier. The information will correlate the release of 5-fluorouracil best fitted with CS polymer (Horo et al., 2019).

1.4.6 DOCETAXEL-LOADED D-A-TOCOPHEROL POLYETHYLENE GLYCOL SUCCINATE 1000 (TPGS) CONJUGATED CHITOSAN (CS) AMALGAMATED TRASTUZUMAB AS ANTICANCER DRUG DELIVERY SYSTEM

Mehata et al., developed trastuzumab loaded docetaxel embedded D-α-tocopherol polyethylene glycol succinate 1000 (TPGS) conjugated CS as an anticancer drug delivery system. The formulation was developed in a stepwise manner as activation of TPGS by succinic anhydride followed by reaction with CS to make complex TPGS-g-CS. Then in the next step, TPGS, TPGS-g-CS (30 mg), docetaxel, and sodium tripolyphosphate were formulated to develop the non-targeted nanoparticulate formulation. Targeted nanoparticle was prepared upon reaction of activated TPGS (15 mg), TPGS (5 mg), TPGS-g-CS (30 mg), docetaxel, sodium tripoly-phosphate and trastuzumab (2.5 mg); finally characterized by particle size distribution, *in vitro* release at pH 7.4, *in vitro* cytotoxicity against SK-BR-3 (Breast Cancer Cell line) and HER-2 (Human Epidermal Growth Factor Receptor 2). Outcomes revealed optimal particle size distribution data, with 78% of entrapment efficiency. As per the dissolution data, the slow release was observed at pH 7.4, and sudden release was found at pH 5.5. IC_{50} value against SKBR-3 showed 1.4793 and 0.2847 μg/ml against non targeted and targeted nanoparticle with significant circulation within the blood. The information stated better delivery of these formulations as anticancer molecules (Mehata et al., 2019).

1.4.7 DECORATED CHITOSAN (CS) NANOPARTICLE WITH TRYPANOCIDAL ACTIVITY

Manuja et al., developed a nanoparticle composed of quinapyramine sulfate (50 mg in sodium tripolyphosphate) and CS [(0.8–1.0) g in 2% acetic acid solution], followed by drug loading and encapsulation efficiency, *in vitro* toxicity measurement against Vero cell line and inhibition of *Trypanosoma evansi*. Observation showed 70.87% of encapsulation capacity, less toxic towards Vero cells. Nanoparticle (with 0.22 mg kg/body weight) showed the presence of parasite up to three days, and in higher concentration, it was abolished. So this decorated CS nanoparticle would work as good trypanocidal agent (Manuja et al., 2018).

1.4.8 COUMARIN LOADED CHITOSAN (CS) NANOPARTICLE AS CONTROLLED DRUG RELEASE CARRIER

Rahimi et al., developed a nanoparticle upon reaction of CS [(1–5)% w/v in glacial acetic acid] and (0.2–1.2)% w/v 8-formyl-7- hydroxy-4-methyl coumarin solution to make a final solution, followed by two divisions. One part was photo-ultraviolet irradiated and mixed irradiated, un-irradiated nanoparticles were centrifuged to obtain the final formulation. The formulation was characterized by scanning electron microscopy, atomic force microscopy, and encapsulation efficiency followed by drug release. It was observed that spherical nanoparticle followed pH-sensitive release kinetics. This observation was correlated with the controlled release behavior of decorated CS nanoparticle for controlled release therapy (Rahimi et al., 2018).

1.4.9 DECORATED CHITOSAN (CS) CONJUGATED DEXTRAN (-SH DERIVATIVE) LOADED MI-RNA NANOPARTICLE FOR CANCER THERAPY

Tekie et al., developed nanoformulation upon the reaction of polyelectrolyte complex (formed upon reaction of miR-145 conjugated thiolated dextran and CS) and aptamer (AS1411) with 1:20 ratio, followed by centrifugation and incubation (Table 1.5). The formulation was characterized by particle size distribution, MTT assay against MCF-7 cell line as principle parameters; which showed particle size (70–272) nm with zeta potential ((–)10–(+)21), with cell toxicity against breast cancer cell line. This observation was associated with nanoparticle development for cancer therapy (Tekie et al., 2018).

1.4.10 SELENOMETHIONINE CONJUGATED CHITOSAN (CS) WITH ZEIN COAT AS ORAL DRUG DELIVERY SYSTEM

Vozza et al., developed nanoformulated particle composed of selenomethionine, CS, and sodium tripolyphosphate with (4:1, 6:1, 8:1) with final molecular weight 30 KDa followed by coated with zein (10 mg/ml in 80% ethanolic solution). The formulation was characterized by encapsulation efficiency, MTS assay against Caco-2 and HepG2 cell lines and in vitro release characteristics in simulated gastric and intestinal fluid. Outcomes revealed that encapsulation efficiency of 85.0% was observed with CS and zein ratio of 2:1; showed no toxicity against Caco-2 cell but minimal toxicity against HepG2 cell line and in simulated intestinal fluid less than 60% drug was released. This information stated that selenomethionine loaded CS coated with zein will improvise an oral drug delivery system (Vozza et al., 2018).

TABLE 1.5   miR-145 PECs Composition

[a]D/Ch	[b]TD-miR (µg/ml)	[c]Ch (µg/ml)
0.2	200	550
1	200	110
5	200	22

[a]D/Ch = dextran to CS molar ratio.
[b]TD-miR (µg/ml) = thiolated dextran–miR-145 conjugates.
[c]Ch (µg/ml) = CS concentration.
Source: Tekie et al., Copyright © 2018 with permission from Elsevier B.V.

1.4.11   CHITOSAN (CS) NANOPARTICLE CONJUGATE WITH POLYSORBATE 80 FOR BRAIN TARGETING DRUG DELIVERY

Ray et al., developed nanoparticle composed of CS coated with polysorbate 80 with ropinirole HCl and polymer ratio of (1:3) with particle size 240nm, 37.62% of entrapment efficiency and zeta potential (–)19.60 mV. They exhibited a higher concentration of ropinirole in the brain than liver, spleen, and kidney. These outcomes confirmed the development of optimum brain targeting drug delivery system (Ray et al., 2018).

1.4.12   SHRIMP DERIVED CHITOSAN (CS) AS MOSQUITO LARVACIDAL ACTIVITY

Anand et al., developed nanoparticle composed of CS (obtained from shrimp cell deacetylation) followed by reaction with sodium tripolyphosphate and

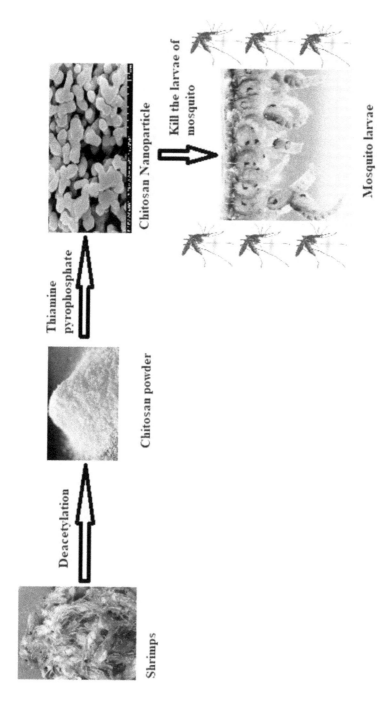

SCHEME 1.2 (See color insert.) Schematic illustration of chitosan extraction from shrimp shells and synthesis of CS NPs from extracted chitosan using TPP as a reducing agent.

analyzed against mosquito larva with different concentration of formulation (20, 40, 60, 80, 100 μL) (Scheme 1.2). The observation came out with 96% of mortality as a larvacidal activity with potential pharmacological efficiency (Anand et al., 2018).

1.5 CONCLUSION AND FUTURE SCOPE

The chapter provides a detailed overview of recent pharmaceutical uses of polysaccharide obtained from marine sources, such as algae, seaweed, coral, and shrimp. Alginate, carrageenan, and CS are common marine polysaccharides. Alginate, carrageenan, and CS were observed with diversified pharmaceutical applications such as controlled release properties, sustained release carrier with anticancer, antivirus, antifungal, anti-inflammatory drugs. They are also applied in wound healing. They have trypanocidal effect in the various pharmaceutical formulations such as hydrogel, aerogel, nanoparticle, vaginal insert, enteric-coated delivery, targeted intestinal delivery, enteric-coated delivery of probiotic bacteria, etc. Nowadays, throughout the world, scientists are focused with their excellence to use a maximum number of natural polymers in the field of drug discovery and development, modified release characteristics, and site-directed delivery system. Due to the biodegradable and biocompatible nature of marine polysaccharides, they are the primary choice of drugs having targeted, controlled, sustainable properties to achieve greater bioavailability, less toxicity, optimum pharmacokinetic, pharmacological behavior.

KEYWORDS

- alginate
- carrageenan
- chitosan
- controlled release
- drug delivery
- nanoformulation

REFERENCES

Abad, L. V., Relleve, L. S., Racadio, C. D. T., Aranilla, C. T., & Rosa, A. M. D., (2013). Antioxidant activity potential of gamma-irradiated carrageenan. *Appl. Radiation Isotopes, 79*, 73–79.

Ahmed, A. B., Adel, M., Karimi, P., & Peidayesh, M., (2014). Pharmaceutical, cosmetic and traditional application of marine carbohydrates. *Adv. Food. Nutr. Res., 73*, 197–220.

Anand, M., Sathyapriya, P., Maruthupandy, M., & Beevi, A. H., (2018). Synthesis of chitosan nanoparticles by TPP and their potential mosquito larvicidal application. *Frontier Lab Medicine, 2*(2), 72–78.

Bugnone, C. A., Ronchetti, S., Manna, L., & Banchero, M., (2018). An emulsification/internal setting technique for the preparation of coated and uncoated hybrid silica/alginate aerogel beads for controlled drug delivery. *The J. Supercritical Fluids, 142*, 1–9.

Catanzano, O., D'Esposito, V., Formisano, P., Boateng, J. S., & Quaglia, F., (2018). Composite alginate-hyaluronan sponges for the delivery of tranexamic acid in post-extractive alveolar wounds. *J. Pharmaceut. Sci., 107*, 654–661.

Chiu, Y. H., Yi-Lin, C. Y. L., Tsai, L. W., Li, T. L., & Wua, C. J., (2012). Prevention of human enterovirus 71 infection by kappa carrageenan. *Antiviral Res., 95*, 128–134.

Craciun, A. M., Tartau, L. M., Pinteala, M., & Marin, L., (2019). Nitrosalicyl-imine-chitosan hydrogels based drug delivery systems for long term sustained release in local therapy. *J. Colloid. Interface Sci., 536*, 196–207.

Dafe, A., Etemadia, H., Zarredar, H., & Mahdavinia, G. R., (2017). Development of novel carboxymethyl cellulose/k-carrageenan blends as an enteric delivery vehicle for probiotic bacteria. *Int. J. Biol. Macromol., 97*, 299–307.

Devi, N., & Dutta, J., (2017). Preparation and characterization of chitosan-bentonite nanocomposite films for wound healing application. *Int. J. Biol. Macromol., 104*(B), 1897–1904.

Freitas, E. D., Vidart, J. M. M., Silva, E. A., Silva, M. G. C., Melissa, G. A., & Vieira, M. G. A., (2018). Development of mucoadhesive sericin/alginate particles loaded with ibuprofen for sustained drug delivery. *Particuology, 41*, 65–73.

Garcia-Astrain, C., & Averous, L., (2018). Synthesis and evaluation of functional alginate hydrogels based on click chemistry for drug delivery applications. *Carbohydr. Polym., 190*, 271–280.

Gomez-Mascaraque, L. G., Llavata-Cabrero, B., Martinez-Sanz, M., Fabra, M. J., & Rubio, A. L., (2018). Self-assembled gelatin-i-carrageenan encapsulation structures for intestinal-targeted release applications. *J. Colloid. Interface Sci., 517*, 113–123.

Horo, H., Das, S., Mandal, B., & Kundu, L. M., (2019). Development of a photoresponsive chitosan conjugated prodrug nano-carrier for controlled delivery of antitumor drug 5-fluorouracil. *Int. J. Biol. Macromolecules, 121*, 1070–1076.

Jalababu, R., Veni, S. S., & Reddy, K. V. N. S., (2018). Synthesis and characterization of dual responsive sodium alginate-g-acryloyl phenylalanine-poly n-isopropyl acrylamide smart hydrogels for the controlled release of anticancer drug. *J. Drug Deli. Sci. Tech., 44*, 190–204.

Joshy, K. S., Susan, M. A., Snigdha, S., Nandakumar, K., Pothen, L. A., & Thomas, S., (2018). Encapsulation of zidovudine in PF-68 coated alginate conjugate nanoparticles for anti-HIV drug delivery. *Int. J. Biol. Macromol., 107*, 929–937.

Karimi, M. H., Gholam, R., & Mahdavinia, G. R., & Massoumi, B., (2018). pH-controlled sunitinib anticancer release from magnetic chitosan nanoparticles crosslinked with κ-carrageenan. *Materials Sci. Eng. C., 91*, 705–714.

Kim, S. K., Ravichandran, Y. D., Khan, S. B., & Kim, Y. T., (2008). Prospective of cosmetic derived from marine organisms. *Biotechnol. Bioprocess Eng., 13*, 511–523.

Lal, M., Lai, M., Ugaonkar, S., Wesenberg, A., Kizima, L., Rodriguez, A., Levendosky, K., Mizenina, O., Fernandez-Romero, J., & Zydowsky, T., (2018). Development of a vaginal fast-dissolving insert combining griffithsin and carrageenan for potential use against sexually transmitted infections. *J. Pharmaceutical Sci., 107*, 2601–2610.

Lee, K. Y., & Mooney, D. J., (2012). Alginate: Properties and biomedical applications. *Prog. Polym. Sci., 37*(1), 106–126.

Li, C., & Li, J. B., (2017). Preparation of chitosan-ferulic acid conjugate: Structure characterization and in the application of pharmaceuticals. *International Journal of Biological Macromolecules, 105*, 1539–1543.

Lim, H. P., Ooi, C. W., Tey, B. T., & Chan, E. S., (2017). Controlled delivery of oral insulin as part using pH-responsive alginate/κ-carrageenan composite hydrogel beads. *Reactive Functional Polym., 120*, 20–29.

Lima, D. S., Tenorio-Neto, E. T., Lima-Tenorio, M. K., Guilherme, M. R., Scariot, D. B., Nakamura, C. V., Muniz, E. C., & Rubira, A. F., (2018). pH-responsive alginate-based hydrogels for protein delivery. *J. Mol. Liq., 262*, 29–36.

Liu, X., Nielsen, L. H., Klodzinska, S. Y., Nielsen, M. H., Qu, H., Christensen, L. P., Rantanen, J., & Yang, M., (2018). Ciprofloxacin-loaded sodium alginate/poly (lactic-co-glycolic acid) electrospun fibrous mats for wound healing. *Euro. J. Pharmaceutic. Biopharmaceutics., 123*, 42–49.

Lopedota, A., Denora, N., Laquintana, V., Annalisa, C. A., Lopalco, A., Tricarico, D., Maqoud, F., Curci, A., Maria, M. M., Forgia, F., Fontana, S., & Franco, M., (2018). Alginate-based hydrogel containing minoxidil/hydroxypropyl-b-cyclodextrin inclusion complex for topical alopecia treatment. *J. Pharmaceut. Sci., 107*, 1046–1054.

Mahdavinia, G. R., Mosallanezhad, A., Soleymani, M., & Sabzi, M., (2017). Magnetic- and pH-responsive κ-carrageenan/chitosan complexes for controlled release of methotrexate anticancer drug. *Int. J. Biol. Macromolecule, 97*, 209–217.

Mahmoud, A. A., Nermeen, A., Elkasabgy, N. A., & Abdelkhalek, A. A., (2018). Design and characterization of emulsified spray dried alginate microparticles as a carrier for the dually acting drug roflumilast. *Euro. J. Pharmaceut. Sci., 122*, 64–76.

Manuja, A., Dilbaghi, N., Kaur, H., Saini, R., Barnela, M., Chopra, M., Manuja, B. K., Kumar, R., Kumar, S., Riyesh, T., Singh, S. K., & Yadav, S. C., (2018). Chitosan quinapyramine sulfate nanoparticles exhibit increased trypanocidal activity in mice. *Nano-Structures, Nano-Objects, 16*, 193–199.

Mehata, A. K., Bharti, S., Singh, P., Viswanadh, M. K., Kumari, L., Agrawal, P., Singh, S., Koch, B., & Muthu, M. S., (2019). Trastuzumab decorated TPGS-g-chitosan nanoparticles for targeted breast cancer therapy. *Colloids Surfaces B: Biointerfaces, 173*, 366–377.

Montaser, R., & Luesch, H., (2011). Marine natural products: A new wave of drugs? *Future. Med. Chem., 3*, 1475–1489.

Mukhopadhyay, P., Maity, S., Mandal, S., Chakraborti, A. S., Prajapati, A. K., & Kundu, P. P., (2018). Preparation, characterization and *in vivo* evaluation of pH-sensitive, safe quercetin-succinylated chitosan-alginate core-shell-corona nanoparticle for diabetes treatment. *Carbohydr. Polym., 182,* 42–51.

Nayak, A. K., Pal, D., & Malakar, J., (2012b). Development, optimization and evaluation of floating beads using natural polysaccharides blend for controlled drug release. *Polymer. Eng. Sci., 53,* 238–250.

Nayak, A., & Pal, D., (2011). Development of pH-sensitive tamarind seed polysaccharide alginate composite beads for controlled diclofenac sodium delivery using response surface methodology. *Int. J. Biol. Macromol., 49,* 784–793.

Nayak, A., & Pal, D., (2012a). Ionotropically-gelled mucoadhesive beads for oral metformin HCl. *Int. J. Pharmaceut. Sci. Rev. Res., 1,* 1–7.

Nayak, A., & Pal, D., (2013). Fenugreek seed gum-alginate mucoadhesive beads of metformin –HCl: Design, optimization and evaluation. *Int. J. Biol. Macromol., 54,* 144–154.

Nayak, A., & Pal, D., (2014a). Ispaghula mucilage-gellan mucoadhesive polysaccharide-gellan mucoadhesive beads for controlled release of metformin HCl. *Carb. Polym., 103,* 41–50.

Nayak, A., & Pal, D., (2014b). Tamarind seed polysaccharide-gellan mucoadhesive beads for controlled release of metformin HCl. *Carb. Polym., 103,* 154–163.

Nayak, A., & Pal, D., (2014c). Development of calcium pectinate-tamarind seed polysaccharide mucoadhesive beads containing metformin HCl. *Carb. Polym., 101,* 220–230.

Nayak, A., & Pal, D., (2015a). *Polymeric Hydrogels as Smart Biomaterials Part of the Springer Series on Polymer and Composite Materials Book Series (SSPCM),* 105–151.

Nayak, A., & Pal, D., (2015b). Plant-derived polymers: Ionically-gelled sustained drug release systems. In: *Encyclopedia of Biomedical Polymers and Polymer Biomaterials* (Vol. 8, pp. 6002–6017). Taylor & Francis. ISBN 9781439898796.

Nayak, A., & Pal, D., (2016). Sterculia gum-based hydrogels for drug delivery applications. In: *Springer Series on Polymer and Composite Materials - Polymeric Hydrogels as Smart Biomaterials.* ISBN: 978-3-319-25320-6.

Nayak, A., Pal, D., & Das, S., (2012c). Calcium pectinate-fenugreek seed mucilage mucoadhesive beads for controlled delivery of metformin HCl. *Carb. Polym., 96,* 349–357.

Necas, J., & Bartosikova, L., (2013). Carrageenan: A review. *Veterinarni. Medicina., 58*(4), 187–205.

Rahimi, S., Khoee, S., & Ghandi, M., (2018). Development of photo and pH dual cross-linked coumarin-containing chitosan nanoparticles for controlled drug release. *Carbohydr. Polym., 201,* 236–245.

Rahmani, V., & Sheardown, H., (2018). Protein-alginate complexes as pH-/ion-sensitive carriers of proteins. *Int. J. Pharmaceutic., 535,* 452–461.

Randy, C. F. C., Tzi, B. N., Jack, H. W., & Wai, Y. C., (2015). Chitosan: An update on potential biomedical and pharmaceutical applications. *Mar. Drugs, 13,* 5156–5186.

Rasmussen, R. S., & Morrissey, M. T., (2007). Marine biotechnology for production of food ingredients. *Adv. Food. Nutr. Res., 52,* 237–292.

Ray, S., Sinha, P., Laha, B., Maiti, S., Bhattacharyya, U. K., & Nayak, A. K., (2018). Polysorbate 80 coated crosslinked chitosan nanoparticles of ropinirole hydrochloride for brain targeting. *J. Drug Delivery Sci. Tech., 48,* 21–29.

Sachan, N. K., Pushkar, S., Jha, A., & Bhattacharya, A., (2009). Sodium alginate: The wonder polymer for controlled drug delivery. *J. Pharmacy. Res., 2*(8), 1191–1199.

Senna, J. P., Barradas, T. N., Cardoso, S., Castiglione, T. C., Serpe, M. J., Kattya, G., De Holanda, E., Silva, K. G. H., & Mansur, C. R. E., (2018). Dual alginate-lipid nanocarriers as oral delivery systems for amphotericin B. Colloid. *Surfaces B: Biointerfaces, 166,* 187–194.

Shao, L., Cao, Y., Li, Z., Hu, W., Li, S., & Lu, L., (2018). Dual responsive intelligent aerogel made from thermo/pH sensitive graft copolymer ALG-g-P(NIPAM-co-NHMAM) for drug controlled release. *Int. J. Biol. Macromole., 114,* 1338–1344.

Shweta, A., Ankita, L., Aakriti, T., Vijay, K., Imran, M., & Anita, K. V., (2015). Versatility of chitosan: A short review. *J. Pharm. Res., 4*(3), 125–134.

Sonawane, R. O., & Patil, S. D., (2018). Fabrication and statistical optimization of starch-κ-carrageenan cross-linked hydrogel composite for extended-release pellets of zaltoprofen. *Int. J. Biol. Macromol., 120*(B), 2324–2334.

Sorasitthiyanukarn, F. N., Muangnoi, C., Bhuket, P. R. N., Rojsitthisak, P., & Rojsitthisak, P., (2018). Chitosan/alginate nanoparticles as a promising approach for oral delivery of curcumin diglutaric acid for cancer treatment. *Materials. Sci. Eng.: C., 93,* 178–190.

Suchaoin, W., Bonengel, S., Hussain, S., Huck, C. W., Ma, B., & Bernkop-Schnuch, A., (2015). Synthesis and *in vitro* evaluation of thiolated carrageenan. *J. Pharmaceutical. Sci., 104,* 2523–2530.

Sukhodub, L. F., Sukhodub, L. B., Litsis, O., & Prylutskyy, Y., (2018). Synthesis and characterization of hydroxyapatite-alginate nanostructured composites for the controlled drug release. *Materials. Chem. Physics., 217,* 228–234.

Summa, M., Russo, D., Penna, I., Margaroli, N., Bayer, I. S., Bandiera, T., Athanassiou, A., & Bertorelli, R., (2018). A biocompatible sodium alginate/povidone-iodine film enhances wound healing. *Euro. J. Pharmaceutics. Biopharmaceutics, 122,* 17–24.

Tekie, F. S. M., Soleimani, M., Zakerian, A., Dinarvand, M., Amini, M., Dinarvand, R., Arefian, E., & Atyabi, F., (2018). Glutathione responsive chitosan-thiolated dextran conjugated miR-145 nanoparticles targeted with AS1411 aptamer for cancer treatment. *Carbohydr. Polym., 201,* 131–140.

Thomas, D., Latha, M. S., & Thomas, K. K., (2018). Synthesis and *in vitro* evaluation of alginate-cellulose nanocrystal hybrid nanoparticles for the controlled oral delivery of rifampicin. *J. Drug. Deli. Sci. Tech., 46,* 392–399.

Tobacman, J. K., (2001). Review of harmful gastrointestinal effects of carrageenan in animal experiments. *Environ. Health Perspect, 109*(10), 983–994.

Volic, M., Pajic-Lijakovic, I., Djordjevic, V., Knezevic-Jugovic, Z., Pecinar, I., Stevanovic-Dajic, Z., Veljovic, D., Hadnadjev, M., & Bugarski, B., (2018). Alginate/soy protein system for essential oil encapsulation with intestinal delivery. *Carbohydr. Polym., 200,* 15–24.

Vozza, G., Danish, M., Byrne, H. J., Frias, J. M., & Ryan, S. M., (2018). Application of Box-Behnken experimental design for the formulation and optimization of selenomethionine-loaded chitosan nanoparticles coated with zein for oral delivery. *Int. J. Pharmaceutics, 551,* 257–269.

Wang, X., Chang, Z., Nie, X., Li, Y., Hu, Z. P., Ma, J., Wang, W., Song, T., Zhou, P., Wang, H., & Yuan, Z., (2019). A conveniently synthesized Pt (IV) conjugated alginate nanoparticle with ligand self-shielded property for targeting treatment of hepatic carcinoma. *Nanomedicine: Nanotech. Biol. Med., 15,* 153–163.

Xie, H., Xiang, C., Li, Y., Wang, L., Zhang, Y., Song, Z., Ma, X., Lu, X., Lei, Q., & Fang, W., (2019). Fabrication of ovalbumin/κ-carrageenan complex nanoparticles as a novel carrier for curcumin delivery. *Food Hydrocolloids, 89*, 111–121.

Yin, Z. C., Wang, Y. L., & Wang, K., (2008). A pH-responsive composite hydrogel beads based on agar and alginate for oral drug delivery. *J. Drug Delivery Sci. Tech., 43*, 12–18.

Yuan, J., Guo, L., Wang, S., Liu, D., Qin, X., Zheng, L., Tian, C., Han, X., Chen, R., & Yin, R., (2018). Preparation of self-assembled nanoparticles of ε-polylysine–sodium alginate: A sustained-release carrier for antigen delivery. *Colloids Surfaces B: Biointerfaces, 171*, 406–441.

CHAPTER 2

Pharmaceutical Applications of Alginates

AMIT KUMAR NAYAK,[1] MOUMITA GHOSH LASKAR,[2]
MOHAMMAD TABISH,[3] MD SAQUIB HASNAIN,[4] and
DILIPKUMAR PAL[5]

[1]Department of Pharmaceutics, Seemanta Institute of Pharmaceutical Sciences, Mayurbhanj – 757086, Odisha, India

[2]Department of Medicine, Karolinska University Hospital, Huddinge, Karolinska Institute, Stockholm, Sweden

[3]Department of Pharmacology, College of Medicine, Saqra University, 1196, Kingdom of Saudi Arabia

[4]Department of Pharmacy, Shri Venkateshwara University, Gajraula, U.P., India

[5]Department of Pharmaceutical Sciences, Guru Ghasidas Vishwavidyalaya, Koni, Bilaspur – 495009, C.G., India

2.1 INTRODUCTION

Substances, which are not pharmacologically active as drugs or pro-drugs but are used in the formulation of pharmaceutical products are called as pharmaceutical excipients (Ansel et al., 2000). These excipients have various important roles in the preparation of pharmaceutical dosages. Pharmaceutical excipients are used to prepare various drug delivery dosage forms to modulate solubility and drug bioavailability for improving the stability of drug candidates to preserve the polymorphic forms, conformation, osmolarity, and pH of the liquid formulations, to improve the binding capacity and disintegration functions (mainly, in tablets) and many more. Excipients are also used to prevent the aggregation or disso-ciation in the dosage forms. There are some pharmaceutical excipients, which have antioxidant properties as well such as cashew gum, okra gum, etc. (Hasnain et al., 2017a, b, 2018a; Nayak et al., 2015, 2018a, b, c). Recent years, a variety of naturally derived polysaccharides represent as the potential pharmaceutical excipients for industrial uses and thus, have already been extensively researched over the comparable synthetic phar-maceutical excipients on the account of easily availability in the natural source, sustainability, biodegradability, biocompatibility, and nonirritant nature (Bera et al., 2015; Das et al., 2013; Guru et al., 2018; Hasnain et al., 2018b; Jena et al., 2018; Nayak, 2016). Besides these benefits, synthetic excipients are not being preferred over the naturally derived polysaccharide based excipients because of the nonrenewable sources, expensive extraction cost, environmental hazards during synthesis and extraction (Nayak and Pal, 2016, 2018; Nayak et al., 2010, 2012, 2018a, d; Sinha et al., 2015a). The current socioeconomic status of the world has elevated the awareness on the use of various natural polysaccharides as potential natural pharmaceutical excipient materials in the formulation of different pharmaceutical drug delivery dosage systems (Das et al., 2014; Hasnain et al., 2010; Nayak and Pal, 2012, 2017a; Pal and Nayak, 2017). Furthermore, the natural polysaccharides are biologically degraded into different physiological metabolites, which make these as a useful group of pharmaceutical excipient candidates for biomedical uses, in particular, drug delivery (Hasnain et al., 2010; Nayak and Pal, 2012; Sinha et al., 2015a, b). During past few decades, various marine-derived polysac-charide excipients have already been invented and investigated as the potential pharmaceutical excipient materials to formulate diverse types of pharmaceutical dosage forms (Malakar et al., 2012; Jana et al., 2014; Nayak and Pal, 2014; Verma et al., 2017). Amongst these, alginates are one of the popular natural polysaccharide groups of alginic acid widely used as pharmaceutical excipients in numerous pharmaceutical applica-tions including sustained drug releasing systems, biomucoadhesive dosage forms, colon-specific drug releasing systems, protein, and peptide delivery systems, gene delivery systems, etc. (Goh et al., 2012; Malakar and Nayak, 2012a; Pal and Nayak, 2015a, 2017; Hasnain and Nayak, 2018). This chapter presents the usual applications of alginates as potential excipients in various kinds of pharmaceutical dosage forms.

2.2 SOURCES AND PRODUCTION OF ALGINATES

Alginates are a high molecular weight natural polymeric group of poly-saccharide nature. From various species of kelp, alginates are extracted (Ertesvag and Valla, 1998). Various brown marine algae, like *Laminaria hyperborea, Ascophyllum nodosum, Macrocystis pyrifera*, etc., are identi-fied as some important primary sources of commercially available alginates (Nayak et al., 2010; Pal and Nayak, 2015a; Sutherland, 1991). However, these are also extracted from some bacterial sources, such as Pseudomonas species, Azotobacter species, etc. But, these are generally restricted to the small-scale uses like research (Draget et al., 2005). Commercially, alginates are mainly isolated by alkaline extraction (Draget et al., 2005; Skjaak-Braek et al., 1986). Due to the presence of insoluble salts containing Ca^{2+}, Mg^{2+}, Na^+, K^+, etc., ions occurred in cell wall- as well as extracellular- matrices of algae. The extraction and subsequent purification procedures of alginates are usually done via the ion exchange procedures. Before the extraction of alginates, the algae is first harvested, dried, and then, treated with a dilute mineral acid to eliminate or degrade some related neutral natured homo-polysaccharides, like, laminarin, fucoidin, etc. (Goh et al., 2012). However, *Macrocystis pyrifera* is processed in wet condition. Alginate is converted to the soluble salt form of sodium from the insoluble protonated form, during the alkaline extraction process through the adding up of sodium carbonate at below pH 10. The alginates are then further purified and transferred to either alginic acid or salts of alginate acid. Alginates, extracted from the natural resources contain some kinds of impurity compounds, such as proteins, other carbohydrates endotoxins, polyphenols, heavy metals, etc. Although presences of low levels of these impurities are acceptable in the foods and beverages, these should be completely removed during their uses in the pharmaceutical products (Sutherland, 1991; The United States Pharmacopoeia, 2000).

2.3 CHEMICAL NATURE OF ALGINATES

Alginates are mainly found in sea-water, and these exist as alginic acid salts of Mg^{2+}, Sr^{2+}, Na^+, etc. (Goh et al., 2012). Chemically, alginates comprise the linear blocks of $(1 \rightarrow 4)$-linked α-L-guluronic acid (G) monomer as well as β-D-mannuronic acid (M) monomer, these monomers are positioned in an irregular wise way of GG, MG, and MM units of varying proportions, having 1,4-glycosidic linkages between these (Figure 2.1). M and G block contents of alginates have different functional role depending upon their length and source (Pal and Nayak, 2017). G-blocks are accountable for the ionic cross-linking interaction with various divalent metallic cations; whereas, the length of G-block affects the physical characteristics of the developed hydrogels (Haug et al., 1966, 1967; George and Abraham, 2006). Composition, molecular weight, and sequences of the copolymer depend on the sources and on the species used for the alginate extraction.

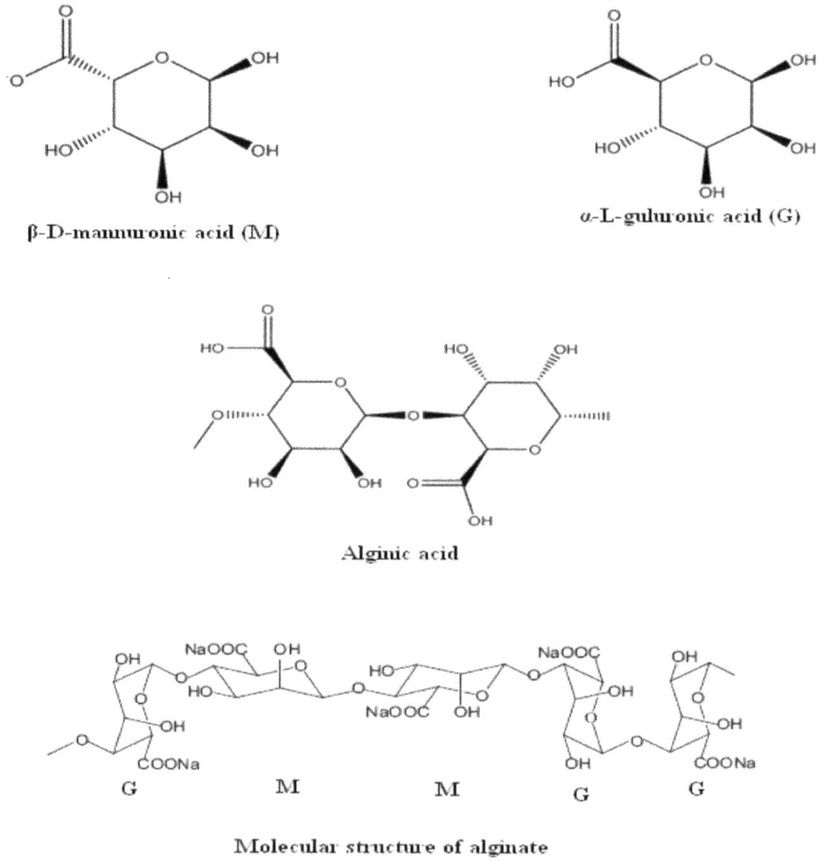

β-D-mannuronic acid (M)

α-L-guluronic acid (G)

Alginic acid

Molecular structure of alginate

FIGURE 2.1 Structure of alginic acid and its monomer units (α-L-Glucoronic Acid, β-D-mannuronic acid).

Additionally, the molecular size decides the viscosity, whereas the guluronic acid residue in the block structure is responsible for affinity toward the cations and also, for the gel-forming properties (Yang et al., 2011; Khalid et al., 2015).

2.4 PROPERTIES OF ALGINATES

2.4.1 SOLUBILITY

Generally, when the –COOH groups are deprotonated, alginates are aqueous soluble in nature. Alginates show insolubility in the aqueous medium, when the –COOH groups are protonated (Goh et al., 2012). In general, sodium alginate is dissolved in the aqueous medium; but, not completely soluble in the solvents of organic nature, such as methyl alcohol, ethyl alcohol, chloroform, ether, etc. (Pawar et al., 2012). The tetra butyl ammonium salts of alginate are very much soluble in the aqueous medium. Even, it is very soluble in ethylene glycol as well as in the solvents of polar aprotic nature. The solubility of alginates in any solvent medium is reported to be influenced by pH, ionic strength, the occurrence of cations as di/trivalent metal cations in the medium directs to the ionic gelation of alginates (Pawar et al., 2011; Shilpa et al., 2003).

2.4.2 VISCOSITY

Different sodium alginate grades are commercially available, forming solutions of different viscosities within 20–400 centipoises (1% aqueous solution at 200°C temperature). On account of the distribution of the polymeric chain lengths, the aqueous solution of sodium alginate is not noticeably Newtonian and also, behaves as a pseudo-plastic natured solution. When dissolved in the pure water, the decreased viscosity of sodium alginate is measured to augment speedily with the dilutions as examined by Focus and Straues (1948). In support of the rheological characteristics of polyelectrolyte solutions depends on the ionic structural features of aqueous solvents to decrease the solution viscosity by reason of the alterations in the polymer conformations (Shilpa et al., 2003). Nevertheless, the rheology of sodium alginate solutions is strongly associated with alginate concentration and with the number and length of the monomer units of the alginate structure (Berth, 1992; Rehm and Valla, 1997). However, guluronic acid rich sodium alginate is found comparatively more soluble in water (Gombotz and Wee, 1998). Furthermore, alginates of the longer segments were reported to have an elevated viscosity pattern in the similar concentrations (Berth, 1992). The molecular weight values of the integrated polymeric segments in the sodium alginate solutions influence the viscosity characteristics of the sodium alginate solution. Interestingly, alginates, which occur from bacterial sources, do not demonstrate such interaction (Jimenez-Escrig and Sanchez-Muniz, 2000; Rehm and Valla, 1997).

2.4.3 CROSS-LINKING AND SOL-GEL TRANSFORMATION BEHAVIOR

Various aqueous solutions of sodium alginate generally go through the sol-gel transformations by the cross-linking processes. In general, the production of cross-linked alginate gels occurs via three different mechanisms (Goh et al., 2012). These are external gelation mechanism, internal gelation mechanism, and mechanism of gelation by cooling. During the construction of larger beads (or microbeads) via the process of external gelation, alginate solutions containing drugs are transported to the cross-linking solutions as the droplets atomized or extruded. Thereafter, the alginate-made gel construction happens by diffusing various cross-linking cations (divalent or trivalent) into the solutions of sodium alginate (Pal and Nayak, 2015a). During the performance of internal gelation, firstly, an insoluble salt of calcium like calcium carbonate ($CaCO_3$) can be supplemented to the solutions of alginate containing drugs and then, the free divalent calcium ions are consequently released via the adjustment of pH by the addition of glacial acetic acid (Chan et al., 2006; Goh et al., 2012). Sequestrant agents, like trisodium citrate, can be introduced for the further maintenances the rate of reaction via the competition with alginate-structure for the free ions of calcium (Goh et al., 2012). The particle sizes, concentration, and pH level of the calcium salt (insoluble) have already been claimed to interfere with rates of the cation releasing. For the gel formation by cooling, alginates, calcium salts, and sequestrants are solubilized at 90°C of temperature. Thereafter, it allows setting via the cooling process (Papageorgiou et al., 1994; Zheng, 1997). The higher thermal energy of the alginate polymeric chains interrupts the alignment of the polymeric structure at the higher temperature, and thus, permanently destabilized any noncovalent type intra-molecular bondings among the neighboring polymeric chains. However, while cooling process at the lesser temperature, the inter-molecular bondings re-establishment amongst the polymeric chains helps to form a tertiary structure and the resulting homogeneous matrix (Papageorgiou et al., 1994; Zheng, 1997).

The polyvalent cations can be replaced as well as preferably be connected to different binding sites of residing monovalent sodium cations in the polyguluronate segments, present in the sodium alginate structure, indicating the arrangement of a cross-linked "egg-box" model (Racovita et al., 2009). Usually, various divalent metal cations (such as, Ca^{2+}, Zn^{2+}, Ba^{2+}, Co^{2+}, Cu^{2+}, Cd^{2+}, Ni^{2+}, Pb^{2+}, Mn^{2+}, etc.) are the divalent cross-linkers, but not the monovalent sodium ions or divalent magnesium ions (Gaumann et al., 2000; Racovita et al., 2009; Rees, 1981). The occurrences of the chelation process at the G-residue of alginate molecular structure consequences in the ionic interaction(s) among the guluronic acid residues, while the van der Waal force within the alginate structure results in the 3-dimensional (3D) gelled-networking. Additionally, throughout the process of cross-linking, the production of the inner sphere ion–alginate complex is thought missing for the divalent magnesium cations (Rees and Welsh, 1977). Braccini et al., (1999) employed a similar kind of model, where they demonstrated higher selectivity of polyguluronate residues in the calcium binding process to exhilarate the character as well as the functioning of the guluronate residues in the gelling process of alginates. The proficient axial–equatorial–axial arrangement in the guluronate segments of the alginate structure promotes the cross-linking process via reciprocally adjusting of oxygen atoms from the neighboring residues inside the cation's coordination sphere (DeRamos et al., 1997). Various cross-linking producing metal cations display a distinct affinity for the alginates; despite of the fact that the rate of the cross-linking process can be affected due to the alginate's chemical compositions (Rees, 1981). The dissimilarities in the binding affinities correlate with the ionic radius, the coordination numbers of the cross-linking metal cations and the occurrence of water for the hydration around the di/trivalent cations for the cross-linking (Sutherland, 1991). It was reported that the positively charged proteins over the isoelectric point might actively be contested with ionic cross-linker ions for -COOH sites available on the polymeric segments of alginate structure (Sherys et al., 1989; Stockwell et al., 1986). In addition, the physical hindrances of the charged matrix in the occurrence of calcium ions in excess concentration may hugely influence the mobility of copper cations into the matrices of calcium alginate (Potter and McFarland, 1996). Throughout the procedure of ionic cross-linking for controlling the gelation of alginates, the sequestrants are frequently utilized. The addition of sequestrants like trisodium citrate works as the retarding reagent by means of the chelation free calcium cations (divalent), which are added to the solutions containing ionic cross-linkers (Draget et al., 1994, 2005; Papageorgiou et al., 1994). The controlled releasing of the calcium ions into the alginate solutions are then assists the development of a homogeneous matrix of alginate (Papageorgiou et al., 1994). It has been established before that the divalent calcium ions cross-link the alginate matrices possibly at the binding sites present within the guluronic (G) segments of the alginate structure (Chapman and Chapman, 1980). The composition of guluronic (G) segments in the alginate structure and the cross-linked extension hugely disturbs the gelled matrix quality (Goh et al., 2012). Alginate matrices consist of a higher ratio of guluronic acid, predictably be rigid as well as brittle in nature, while the alginates containing low L-guluronic acid produce matrices with higher elasticity (Shilpa et al., 2003). Alginate pellets that contain more than 70% guluronic acid and the guluronic blocks length of > 15 have proven to produce lesser shrinkage with the stronger mechanical power. Matrices, which are comparatively physically stronger together with the higher porosity, are considered as potentially advantageous for the use in immobilizations of the living cells (Goh et al., 2012; Shilpa et al., 2003). In contrast, the gel strength behavior aroused to be unaltered with the molecular weight of > 200 kDa (Shilpa et al., 2003). Moreover, the following order of polymer segments (MG > MM > GG) is important to induce the flexibility of alginate matrices (Smidsroed et al., 1973).

2.4.4   BIOCOMPATIBILITY AND IMMUNOGENICITY

Biocompatibilities, as well as immunogenicity of the polymeric systems, are the two vital concerns for the successful applications as carriers for the drug delivery applications. Biocompatibility of alginates has already been investigated via injecting the calcium alginate into the kidneys of rodent animals. The outcomes from these studies have established that calcium alginate is biocompatible in nature (Gombotz and Wee, 1998; Shilpa et al., 2003). Calcium alginate gels are found to be non-toxic to cells, and therefore, these are recognized as an acceptable biopolymeric material for the uses in the pharmaceutical drug delivery formulations. Several researches have scrutinized the higher levels of safety issues of the sodium alginate in various food products. Different allergy tastings of sodium alginate and other alginates (like calcium alginate, zinc alginate, aluminum alginate, etc.) have also been performed, and the results of these tests have suggested that these alginates are non-allergic in nature (Dusel et al., 1986).

Most of the research indicates that the presence of mitogenic contaminant materials occurred in the alginates raw materials is the important most contributors to the immunogenicity (George and Abraham, 2006; Shilpa et al., 2003). Both the commercial alginate of research grade and alginate of ultrapure grade are examined for their levels of endotoxin and their capacity to activate lymphocytes. Most common side effects are cytokine releasing, inflammatory responses, etc. (George and Abraham, 2006). Therefore, it has been suggested that the alginate of ultrapure grade with a higher content of α-L-guluronic acid and lower content of β-D-mannuronic acid can be employed for the *in vivo* studies to avoid the inflammatory responses (Gombotz and Wee, 1998).

2.4.5 BIOADHESION

Bioadhesion property of any biomaterials is commonly described as the adhesion or attachment with two different surfaces that one of these should be a biological substratum (Bernkop-Schnürch et al., 2001). Mucoadhesion is called when one of these surfaces is a mucous layer (Park, 1989). Alginate consists of the bioadhesive characteristics and extensively exploited in the development of different mucoadhesive drug deliveries (Bernkop-Schnürch et al., 2001). Various mucoadhesive drug deliveries are capable of producing the enhancement of drug residence period at the application sites of the activity or resorption sites. Thus, the enhancement of effectiveness, as well as bioavailability of the dosage forms, can be obtained (Bernkop-Schnürch et al., 2001).

2.5 PHARMACEUTICAL APPLICATIONS

Alginates possess an extensive potential as the excipient in different phar-maceutical dosage forms like emulsions, suspensions, gels, tablets, capsules, buccal patches, oral particulates, etc., because of its favorable characteristics, including thickening, gel forming, stabilizing, and biocompatibility (Nahar et al., 2017). The alginate molecules can be tailor-made for the numerous uses apart from the common applications as pharmaceutical excipients. These tailor-made alginate materials are also employed to formulate various improved and smart drug delivery systems (Yang et al., 2011).

2.5.1   *THICKENING, EMULSIFYING AGENTS, AND STABILIZING*

Alginates are exploited as a thickening, stabilizing, and emulsifying agents for producing various food products (Brownlee et al., 2005). Sodium alginate is employed as an effective hydrocolloid producing material in various foods and pharmaceutical products as it improves the organoleptic characteristics, moisture retention capacity and the formulation texture (Paraskevopoulou et al., 1994, 2005).

2.5.2   *GELLING AGENTS*

Alginates are capable of forming gels at the lower temperatures for the food applications, which get damaged under the lower temperatures (like meat preparations, fruits, vegetables, etc.). Alginates have been employed for restructured applications as an effective gelling agent in the foods and pharmaceuticals (Cong-Gui et al., 2006). Recent years, a variety of alginate hydrogels have already been fabricated by different techniques based on the physical as well as chemical cross-linking procedures. The structural similarities of these alginate hydrogels to the extracellular matrices of the living tissue allowing a wider applications in delivery of numerous bioactive materials (like small molecular and macromolecular drugs, proteins, and peptides, enzymes, hormones, cells, etc.), wound healing, cell transplantation, etc. (Yang et al., 2011). Alginate hydrogels are also being exploited for the delivery of drugs over a prolonged period (Pal and Nayak, 2015a, 2017). The alginate hydrogels are releasing drugs in a controlled pattern dependent on the types of cross-linking agents and techniques of cross-linking (Racovita et al., 2009).

2.5.3   USES IN CAPSULES AND TABLETS FOR ORAL DRUG DELIVERY

Alginates have been exploited as excipients in the formulation of capsules and tablets for oral administrations of various kinds of drugs (Malakar and Nayak, 2012; Sriamornsak and Sungthongjeen, 2007; Veski and Marvola, 1993). In the pharmaceutical capsules and tablets, alginates of various grades are employed as diluents, matrix-formers, release retardants, pH-sensitive agents, bioadhesive agents, etc. In a research by Veski and Marvola (1993), sodium alginate was investigated as a pharmaceutical diluent in the hard gelatine capsules, where ibuprofen releasing was studied. The research clearly exhibited the potential of the use of algi-nate as capsule diluents. In a research by Ojantakanen et al., (1993), the ibuprofen bioavailability from the hard gelatin capsules made of sodium alginate and/or sucralfate was studied. Veski et al., (1994) have researched the biopharmaceutical potential of pseudoephedrine HCl capsules made of various sodium alginate grades.

Alginates also possess a longer as well as an established background of their exploitations in the formulation of numerous hydrophilic matrices for the controlled drug releasing (Sriamornsak and Sungthongjeen, 2007). Sodium alginate is utilized individually or with other hydrophilic polymeric materials to provide sustained releasing of a variety of neutral, acidic, and basic drug candidates. Sodium alginate is capable of forming porous natured insoluble alginic acid in the gastric fluid, which minimizes the releasing of active ingredients in the stomach. This property is useful to provide protection to the acid-responsive drug candidates or prevents the burst releasing of highly aqueous soluble drug candidates in the stomach. Sodium alginate disintegrates for the slower drug-releasing, in the intestinal pH milieu. When sodium alginate was exploited along with various pH-responsive hydrophilic polymers, like hydroxylpropyl meth-ylcellulose, carboxymethyl cellulose, etc., the release profile is found to be further modified (Malakar and Nayak, 2012b). In the already reported literature, some studies on the theophylline releasing from the capsules containing alginates matrices were reported by Sriamornsak and Sungth-ongjeen (2007). They studied the theophylline releasing characteristics from the hard gelatin capsules of alginate with or without calcium acetate. In this work, calcium ions acted as the ionic cross-linker to produce calcium alginate gel and thereby, the theophylline releasing from these hard gelatin capsules was modified to attain prolonged theophylline releasing capability over a longer period (Sriamornsak and Sungthongjeen, 2007).

Liew et al., (2009) evaluated the potential of sodium alginate as the drug-releasing modifier in the matrix tablets. In this study, seventeen grades of sodium alginates of various compositions, viscosities, particle sizing, etc., were employed for the manufacturing of matrix tablets, which were then used for screening the issues influencing releasing behaviors of drugs from these alginate-based polymeric matrices. The study has shown that the particle sizing is the important issue in the preliminary formation of alginic acid made the gel-barrier as this influenced the rate of burst releasing of drugs, whereas, the higher viscosity of alginate delayed the rate of drug releasing in the buffered phase. However, it augmented the releasing rate of the drugs in the acidic milieu. Moreover, the high M-alginate may be further beneficial than the high G-alginate in producing sustained drug releasing capability. The influence of increased alginate concentrations was found to be superior with the larger sized particles of alginate. Furthermore, the study performed by Liew et al., (2009) has indicated that the matrices of sodium alginate are capable of controlling the drug releasing for 8 hours, minimum, for the highly aqueous soluble drug candidates in the occurrence of an aqueous soluble excipient.

Pornsak et al., (2007) reported the influence of alginate grades on the alginate-based matrix system. In the acidic pH, the hydrated alginate layer was found to be lowered viscous as well as lesser adhesivity. In addition, the texture characteristics of the alginate-based matrix system were found rubbery in nature. This occurrence was probably owing to the conversion of sodium alginate to alginic acid. The alginate erosion was observed as less in the acidic milieu and then, in the neutral medium. This may be because of the development of tough rubbery gel. The releasing of drugs from the alginate matrices was found to be dependent on the dissolution medium pH. In the acidic milieu, the alginate system was found to be fit into the Korsmeyer-Peppas model, which highlighted the drug releasing followed by diffusion as well as erosion dependent mechanisms.

Furthermore, the drug releasing in the neutral medium followed the zero order drug releasing kinetics (Pornsak et al., 2007). The physicochemical properties of drugs are also a significant parameter for the drug releasing mechanism. The releasing of hydrophilic drug candidates follows the diffusion mechanism, whereas the releasing of hydrophobic drug candidates depend on the alginate matrix erosion. Additionally, the porous nature of the alginate-based matrix influences the releasing behaviors of various kinds of drugs. The alginate concentration is directly proportional to the porosity of matrix structure enhancing concomitantly with the increment of alginate concentrations. This occurrence induces the pore structures to the alginate-based matrices and leads to diffusion regulated drug releasing. Reduction in the pore size causes a higher degree of swelling as well as the shrinkage of alginate matrices (Sriamornsak and Sungthongjeen, 2007). Various hydrophilic matrix systems used in pharmaceutical tablets and capsules (already reported) are presented in Table 2.1.

2.5.4   USES IN ORAL PARTICULATES (MICROPARTICLES/BEADS) FOR DRUG DELIVERY

Recent years, several oral particulates, such as microparticles, beads, etc., made of alginates are being investigated, and these oral particulates have been attracted increasing attention by facilitating better drug delivery (Efentakis and Koutlis, 2001; Nayak and Pal, 2017b; Pal and Nayak, 2017). A variety of particulate carriers for the releasing of drug delivery show several key benefits over the single-unit systems like tablets or capsules, since these have been shown the decreasing of inter- as well as intra-subject variabilities in the absorptions of drugs and for the minimization of probability of dose-dumping chances (Nayak and Pal, 2016; Nayak, 2016). To deliver the total doses (recommended), these oral particulates made of alginates are usually capsulated within the capsules, filled within the sachets, and compressed into the tablets (Pal and Nayak, 2015a). These systems are able to mix up with the gastrointestinal fluid. These are also capable of distributing over the longer gastrointestinal region. This decreases the probabilities of impairment of dosage by reducing the malfunction by some units with facilitating more predictable drug delivery capability (Nayak, 2016). All these reported alginate-based oral particulate systems were prepared by a variety of techniques such as covalent cross-linking technique, ionic-gelation technique, combination of ionic-gelation/covalent cross-linking techniques, ionic emulsion-gelation technique and

TABLE 2.1 Some Examples of Alginate-Based Hydrophilic Matrix Systems Used in Pharmaceutical Tablets and Capsules

Various hydrophilic matrix tablets and capsules made of alginates	References
Hard gelatine capsules containing ibuprofen (alginate used as diluent)	Veski and Marvola, 1993
Pseudoephedrine HCl capsules containing different grades of sodium alginate	Veski et al., 1994
Ibuprofen releasing from hard gelatin capsules containing sucralfate or sodium alginate	Ojantakanen et al., 1993
Theophylline releasing from hard gelatin capsules containing various grades of alginate with or without calcium acetate	Sriamornsak and Sungthongjeen, 2007
Theophylline releasing from hard gelatin capsules containing various hydrophilic polymers (HPMC K4M)	Malakar and Nayak, 2012b
Slow-release alginate tablets	Klaudianos, 1972
Alginate matrix tablets prepared by wet granulation	Mandal et al., 2009
Directly compressed alginate tablets for sustained release	Holte et al., 2003
Prolonged-release press-coated ibuprofen tablets containing sodium alginate	Sirkiä et al., 1994
Alginate-based matrix tablets	Sriamornsak et al., 2007
In situ cross-linked alginate matrix tablets for sustained salbutamol sulfate release	Malakar et al., 2014a
Matrix tablets based on carbopol (R) 971P and low-viscosity sodium alginate for pH-independent controlled drug release	Al-Zoubi et al., 2011
Chitosan-sodium alginate matrix tablets	Li et al., 2015
Ca^{2+} ion cross-linked interpenetrating network matrix tablets of polyacrylamide-grafted-sodium alginate and sodium alginate for sustained release of diltiazem HCl	Mandal et al., 2010

extrusion-spheronization technique (Hua et al., 2010; Malakar et al., 2012; Pal and Nayak, 2015b). Some examples of alginate-based particulate systems for oral drug delivery are presented in Table 2.2.

2.5.5   USES IN NANOPARTICLES FOR DRUG DELIVERY

Alginates have already been exploited for the formulation of a variety of nanocarriers for the improvement of bioavailability, targeting of different anticancer drugs to the cancerous cells since last few years. Ahmad et al., (2006) developed nanoparticles made of alginate formulated by using the ionic-gelation technique. These ionically gelled antituberculosis drugs loaded alginate nanoparticles showed improved oral bioavailability in comparison with that of the orally administered free antitubercular drugs. Additionally, these antitubercular drugs encapsulated ionically gelled alginate nanoparticles exhibited a complete clearance of tuberculosis-causing bacteria from various vital organs within the 15 days; whereas the oral administration of free antitubercular drugs had the 45 conventional doses to exhibit the same antitubercular action. In an investigation, Bhowmik et al., (2006) formulated alginate nanocapsules of testosterone via the *in situ* nano-emulsification polymer cross-linking technique and also, studied the sustained releasing potential of the encapsulated testosterone, *in vitro*. These testosterone-loaded alginate nanocapsules exhibited slower release of encapsulated testosterone over a longer period. The same group tested the *in vivo* pharmacokinetic potential of testosterone-loaded nanocapsules in the rats (Jana et al., 2015b). The *in vivo* pharmacokinetic outcome of the testosterone-loaded alginate nanocapsules in the rats demonstrated an improved bioavailability as compared to that of the administration of testosterone injection (commercial) and pure testosterone.

Alginate nanoparticles are also being exploited for multiple drug delivery systems for various hydrophilic as well as hydrophobic drugs. In a study, Jin et al., (2014) formulated calcium carbonate-alginate nanoparticles for the use in combination therapeutics by means of the coprecipitation technique in the aqueous solution. The calcium carbonate-alginate nanoparticles loaded with the dual drugs exhibited significantly improved cellular uptake as well as the nuclear localization in comparison with to the calcium carbonate-alginate nanoparticles loaded with the single drug. Additionally, the cellular inhibitory action in the tumor cells was found

TABLE 2.2   Some Examples of Alginate-Based Particulate Systems for Oral Drug Delivery

Various alginate-based particulate systems for oral drug delivery	References
Calcium alginate microparticles for oral administration	Takka and Acartürk, 1999
Sulindac loaded alginate beads	Yegin et al., 2007
Furosemide-loaded alginate microspheres	Das and Senapati, 2008
Calcium alginate beads containing ampicillin	Torre et al., 1998
Alginate beads of gliclazide	Al-Kassas et al., 2007
Calcium alginate beads containing metronidazole	Patel et al., 2006
Barium chloride cross-linked alginate based diclofenac sodium beads	Morshad et al., 2012
Alginate-diltiazem hydrochloride beads	El-Kamel et al., 2003
Alginate-pectinate bioadhesive microspheres of aceclofenac	Chakraborty et al., 2010
Gliclazide-loaded alginate-methyl cellulose mucoadhesive microcapsules	Pal and Nayak, 2011
Nifedipine-loaded pH-sensitive alginate-chitosan hydrogel beads	Dai et al., 2008
Alginate/chitosan particulate systems for sodium diclofenac release	Gonzalez-Rodriguez et al., 2002
Calcium alginate-chitosan beads of verapamil	Pasparakis and Bouropoulos, 2006
pH-sensitive alginate-chitosan hydrogel beads of carvedilol	Meng et al., 2011
Chitosan-alginate multilayer beads for controlled release of ampicillin	Anal and Stevens, 2005
Alginate/chitosan microparticles containing prednisolone	Wittaya-Areekul et al., 2006
Soluble starch-blended Ca^{2+}-Zn^{2+}-alginate composites-based microparticles of aceclofenac	Nayak et al., 2018
Alginate/ispaghula husk beads of gliclazide	Nayak et al., 2010
Plantago ovata F. Mucilage-alginate mucoadhesive beads for controlled release of glibenclamide	Nayak et al., 2013c
Zinc alginate/okra gum beads of diclofenac sodium	Sinha et al., 2015a
Calcium alginate/okra gum mucoadhesive beads of gliclazide	Sinha et al., 2015b

TABLE 2.2 (*Continued*)

Various alginate-based particulate systems for oral drug delivery	References
Metformin HCl-loaded tamarind seed polysaccharide-alginate beads	Nayak and Pal, 2013a; Nayak et al., 2016
Tamarind seed polysaccharide-alginate mucoadhesive microspheres for oral gliclazide delivery	Pal and Nayak, 2012
pH-sensitive tamarind seed polysaccharide-alginate composite beads for controlled diclofenac sodium delivery	Nayak and Pal, 2011
Calcium alginate/gum Arabic beads of glibenclamide	Nayak et al., 2012b
Linseed polysaccharide-alginate beads of diclofenac sodium	Hasnain et al., 2018b
Fenugreek seed mucilage-alginate mucoadhesive beads of metformin HCl	Nayak et al., 2013a
Jackfruit seed starch-alginate beads containing pioglitazone	Nayak et al., 2013b
Jackfruit seed starch-alginate mucoadhesive beads of metformin HCl	Nayak and Pal, 2013b
Zinc alginate-carboxymethyl cashew gum microbeads of isoxsuprine HCl	Das et al., 2014
Polyethyleneimine-treated or -untreated calcium alginate beads loaded with propranolol-resign complex	Halder et al., 2005
pH-sensitive sodium alginate/poly(vinyl alcohol) hydrogel beads prepared by combined Ca^{2+} crosslinking and freeze-thawing cycles for controlled release diclofenac sodium	Hua et al., 2010
Aceclofenac-loaded unsaturated esterified alginate/gellan gum microspheres	Jana et al., 2013
Alginate-based bipolymeric-nanobioceramic composite matrices	Hasnain et al., 2016
Modified starch (cationized)-alginate beads containing aceclofenac	Malakar et al., 2013a
Potato starch-blended alginate beads for prolonged release of tolbutamide	Malakar et al., 2013b
Novel alginate hydrogel core-shell systems for combination delivery of ranitidine HCl and aceclofenac	Jana et al., 2015a

significantly improved with the dual drug loaded calcium carbonate-alginate nanoparticles. This designated that the dual drug loaded calcium carbonate-alginate nanoparticles loaded can be explored as multiple drug delivery systems against drug-resistant diseases. Some examples of alginate-based nanoparticle systems for drug delivery are presented in Table 2.3.

2.5.6   USES IN GASTRORETENTIVE DRUG DELIVERY SYSTEMS

Alginates are also being employed for the designing of gastroretentive floating delivery formulations of different drugs for achieving the prolong action at the upper gastrointestinal tract (GIT) (Ma et al., 2008). Numerous alginate-based floating beads or microparticles suffered from the burst drug releasing related problems, which can be overcome by the incorporation or reinforcement of the supplementary additive excipients. The floatation behavior (i.e., buoyancy) in these alginate beads and microparticles can be attained via the entrapment of different oils of low density (for examples: liquid paraffin, castor oil, mentha oil, linseed oil, olive oil, groundnut oil, sunflower oil, etc.) (Chaudhuri and Kar, 2005; Malakar et al., 2012; Nayak et al., 2013c). Furthermore, the buoyancy in the alginate-based beads and microparticles can be achieved by using the gas generation technique (Singh et al., 2010). In these beads and microparticles, carbon dioxide or other inert gases producing substances are introduced. In general, calcium carbonate, sodium bicarbonate, and citric acid are employed for gas generation within these beads and microparticles. The buoyancy in these alginate-based particles (beads and microparticles) can be improved via the incorporation of various foam materials (Yao et al., 2012) and magnesium stearate (low-density substance) (Malakar and Nayak, 2012a). Various low-density porous substances like calcium silicate can be rein-forced within the alginate beads and microparticles to impart the buoyancy (Javadzadeh et al., 2010; Ishak et al., 2007). Furthermore, the hydrophilic behavior enhances the dispersal of drug-loaded powders in the aqueous solution of sodium alginate. Some reported alginate-based floating beads for the use in gastroretentive drug delivery is presented in Table 2.4.

In a research by Malakar et al., (2014b), salbutamol sulfate encap-sulated floating beads (oil entrapped) made of alginate, hydroxypropyl methylcellulose (HPMC) and potato starch was developed. These floating beads of salbutamol sulfate were filled within hard gelatine capsules. These hard gelatine capsules containing oil entrapped floating beads of

TABLE 2.3 Some Examples of Alginate-Based Nanoparticle Systems for Drug Delivery

Alginate-based nanoparticle systems for drug delivery	References
Sodium alginate nanospheres of methotrexate	Santhi et al., 2005
Metformin-loaded alginate nanoparticles	Kumar et al., 2017
Inhalable alginate nanoparticles as antitubercular drug carriers	Ahmad et al., 2005
Alginate nanoparticle-encapsulated azole antifungal and antitubercular drugs	Ahmad et al., 2007
Alginate nanoparticle for the targeted delivery of curcumin	Anirudhan et al., 2017
Alginate nanoparticles containing curcumin and resveratrol	Saralkar and Dash, 2017
Thiolated alginate-albumin nanoparticles stabilized by disulfide bonds	Martinez et al., 2011
Disulfide cross-linked nanospheres from sodium alginate derivative for inflammatory bowel disease	Chang et al., 2012
Alginic acid nanoparticles prepared through counter ion complexation	Cheng et al., 2012
Doxorubicin-loaded glycyrrhetinic acid-modified alginate nanoparticles for liver tumor chemotherapy	Zhang et al., 2012
Exemestane loaded alginate nanoparticles for cancer treatment	Jayapal and Dhanaraj, 2017
Brain-targeted intranasal alginate nanoparticles for treatment of depression	Haque et al., 2014
Glutathione-responsive nanoparticles from a sodium alginate derivative for selective release of doxorubicin in tumor cells	Gao et al., 2017
Alginate-chitosan-pluronic composite nanoparticles for curcumin delivery to cancer cells	Das et al., 2010
Magnetically driven alginate nanoparticles for targeted drug delivery	Ciofani et al., 2008
Functionalized magnetic iron oxide/alginate core-shell nanoparticles for targeting hyperthermia	Liao et al., 2015

TABLE 2.4 Some Alginate-Based Floating Beads for Gastroretentive Drug Delivery

Floating alginate beads for gastroretentive drug delivery	References
Cloxacillin loaded oil entrapped alginate beads	Malakar et al., 2012
Oil entrapped alginate gel beads containing metformin HCl	Chaudhuri and Kar, 2005
Domperidone loaded mineral oil entrapped alginate gel buoyant beads	Singh et al., 2011
Alginate-sterculia gum gel-coated oil-entrapped alginate beads for gastroretentive risperidone delivery	Bera et al., 2015a
Alginate gel-coated oil-entrapped alginate–tamarind gum–magnesium stearate buoyant beads of risperidone	Bera et al., 2015b
Gastroretentive floating sterculia-alginate beads for use in antiulcer drug delivery	Singh et al., 2010
Ibuprofen-loaded buoyant alginate beads	Malakar and Nayak, 2012a
Oil-entrapped sterculia gum-alginate buoyant beads of aceclofenac	Guru et al., 2013
Metronidazole floating alginate beads based on calcium silicate and gas forming agents	Javadzadeh et al., 2010
Metronidazole-loaded alginate beads containing calcium silicate	Ishak et al., 2007
Floating alginate microparticles based on low-density foam powder	Streubel et al., 2002
Floating alginate/poloxamer inner-porous beads using foam solution	Yao et al., 2012

salbutamol sulfate were evaluated for the *in vitro* performances and these developed floating systems were tested for X-ray imaging, *in vivo*. The results of *in vitro-in vivo* evaluations demonstrated a good gastroretentive drug-releasing efficacy. Sodium alginate is also being employed to prepare gastroretentive floating tablets. In a research, Babu et al., (2011) formulated gastroretentive-floating tablets of metronidazole on the basis of the mechanism of gas generation through a chemical reaction. To prepare these floating tablets, HPMC (K 15M), sodium alginate, dicalcium phosphate, citric acid, sodium bicarbonate, light magnesium carbonate, and magnesium stearate were used. These gastroretentive-floating tablets of metronidazole exhibited good buoyancy and desired drug releasing in the gastric pH.

2.5.7   USES IN PROTEIN DELIVERY

Delivery of proteins has rapidly been growing with respect to the commercial market (Nayak, 2010). Alginates are the useful candidate for the encapsulation as well as delivery of different kinds of peptides and proteins (Chan and Neufeld, 2010; George and Abraham, 2006; Goh et al., 2012). Alginates are employed to encapsulate a variety of proteins with maintaining their three-dimensional structure, which minimizes the denaturation of protein structure as well as facilitates the protection against the degradation of proteins in the external biological milieu. A variety of approaches such as DNA delivery, encapsulation of enzymes and hormones, etc., have been already researched for controlling the release of proteins and peptides from various alginate-based systems (Chan and Neufeld, 2010; Jain and Amiji, 2012; Nesamony et al., 2012). Some reported alginate-based systems for protein delivery is presented in Table 2.5.

2.6   CONCLUSIONS

A range of marine-derived natural polysaccharide materials has been already employed as useful and effective pharmaceutical excipients to prepare diverse kinds of pharmaceutical drug delivery systems. Amongst the marine-derived natural polysaccharides, alginates (a group of alginic acid salts) have gained popularity for their usefulness as effective pharmaceutical excipients in numerous pharmaceutical applications including sustained release systems, colonic drug releasing systems, bioadhesive

TABLE 2.5   Some Alginate-Based Systems for Protein Delivery

Alginate-based systems for protein delivery	References
Calcium alginate nanoparticles for protein delivery	Nesamony et al., 2012
Core-shell hybrid nanoparticles based on the layer-by-layer assembly of alginate and chitosan on liposomes for protein release	Haidar et al., 2008
Alginate-chitosan coacervate microcapsules for protein release	Vandenberg et al., 2001
Extended releasing of high pI proteins from alginate microspheres	Wells and Sheardown, 2007
Controlled release of albumin from chitosan-alginate microcapsules	Polk et al., 1994
Silk coatings on PLGA and alginate microspheres for protein delivery	Wang et al., 2007
Tuneable semi-synthetic network alginate for protein release	Chan and Neufeld, 2010
Oxidized sodium alginate-graft poly (2-dimethylamino) ethyl methacrylate)gel beads for BSA releasing	Gao et al., 2009
Physically crosslinked alginate/N, O-carboxymethyl chitosan hydrogels for oral delivery of protein drugs	Lin et al., 2005
Tuftsin-modified alginate nanoparticles as a noncondensing macrophage-targeted DNA delivery system	Jain and Amiji, 2012
Alginate modified nanostructured calcium carbonate with enhanced delivery efficiency for gene delivery	Zhao et al., 2012

systems, protein, and peptide delivery systems, gene delivery systems, etc. The current chapter demonstrates the uses of alginates as effective pharmaceutical excipients in the preparation of different pharmaceutical dosage systems like emulsions, gels, tablets, capsules, buccal patches, oral particulates, nanoparticles, etc.

2.7   SUMMARY

Alginates, a group of alginic acid salts, are the naturally derived polysaccharides mainly extracted from various species of kelp. The chemical structure of alginates comprises the linear blocks of $(1 \rightarrow 4)$-linked α-L-guluronic acid (G) monomer as well as β-D-mannuronic acid (M) monomer. These linear blocks are characterized as anionic copolymer comprised of α-L-guluronic acid (G unit) as well as β-D-mannuronic acid (M unit) arranged in the irregular arrangement of various proportions of GG, MG, and MM units, having 1, 4-glycosidic linkages among the monomers. Because of their favorable physic-chemical properties like solubility, viscosity, cross-linking, sol-gel transformation, etc., and improved biological properties like immunogenicity, biocompatibility, bioadhesion, etc., alginates are widely employed as useful and effective excipients in different pharmaceutical drug delivery systems such as emulsions, gels, capsules, tablets, buccal patches, oral particulates, nanoparticles, etc. Recent years, various tailor-made alginate materials are being synthesized and exploited to formulate various improved and smart drug delivery systems. The current chapter presents the most recent uses of alginates as potential excipients in various pharmaceutical dosage forms.

KEYWORDS

- **drug delivery**
- **pharmaceutical excipients**
- **sodium alginate**

REFERENCES

Ahmad, Z., Pandey, R., Sharma, S., & Khuller, G. K., (2006). Alginate nanoparticles as antituberculosis drug carriers: Formulation development, pharmacokinetics and therapeutic potential. *Indian J. Chest. Dis. Allied. Sci.*, *48*, 171–176.

Ahmad, Z., Sharma, S., & Khuller, G. K., (2005). Inhalable alginate nanoparticles as antitubercular drug carriers against experimental tuberculosis. *Int. J. Antimicrob. Agents*, *26*, 298–303.

Ahmad, Z., Sharma, S., & Khuller, G. K., (2007). Chemotherapeutic evaluation of alginate nanoparticle-encapsulated azole antifungal and antitubercular drugs against murine tuberculosis. *Nanomedicine*, *3*, 239–243.

Al-Kassas, R., Al-Gohary, O. M. N., & Al-Fadhel, M. M., (2007). Controlling of systemic absorption of gliclazide through incorporation into alginate beads. *Int. J. Pharm.*, *341*, 230–237.

Al-Zoubi, N. M., AlKhatib, H. S., & Obeidat, W. M., (2011). Evaluation of hydrophilic matrix tablets based on carbopol (R) 971P and low-viscosity sodium alginate for pH-independent controlled drug release. *Drug Dev. Ind. Pharm.*, *37*, 798–808.

Anal, A. K., & Stevens, W. F., (2005). Chitosan-alginate multilayer beads for controlled release of ampicillin. *Int. J. Pharm.*, *290*, 45–54.

Anirudhan, T. S., Anila, M. M., & Franklin, S., (2017). Synthesis characterization and biological evaluation of alginate nanoparticle for the targeted delivery of curcumin. *Mater. Sci. Eng. C. Mater. Biol. Appl.*, *78*, 1125–1134.

Ansel, H. C., Popovich, N. G., & Allen, Jr. L. V., (2000). Farmacotécnica: Formas farmacêuticas e sistemas de liberação de fármacos. *São Paulo: Editorial Premier*.

Babu, G. D., Chandra, S. R., Devi, A. S., Reddy, B. V. V., & Latha, N. S., (2011). Formulation and evaluation of novel effervescent metronidazole floating tablets. *Int. J. Res. Pharm. Biomed. Sci.*, *2*, 1657–1662.

Bera, H., Boddupalli, S., Nandikonda, S., Kumar, S., & Nayak, A. K., (2015b). Alginate gel-coated oil-entrapped alginate–tamarind gum–magnesium stearate buoyant beads of risperidone. *Int. J. Biol. Macromol.*, *78*, 102–111.

Bera, H., Kandukuri, S. G., Nayak, A. K., & Boddupalli, S., (2015a). Alginate-sterculia gum gel-coated oil-entrapped alginate beads for gastroretentive risperidone delivery. *Carbohydr. Polym.*, *120*, 74–84.

Bernkop-Schnürch, A., Kast, C. E., & Richter, M. F., (2001). Improvement in the mucoadhesive properties of alginate by the covalent attachment of cysteine. *J. Control. Release*, *71*, 277–285.

Berth, G., (1992). Methodical aspects of characterization of alginate and pectate by light scattering and viscometry coupled with GPC. *Carbohydr. Polym.*, *19*, 1–9.

Bhowmik, B. B., Sa, B., & Mukherjee, A., (2006). Preparation and *in vitro* characterization of slow release testosterone nanocapsules in alginates. *Acta Pharm.*, *56*, 417–429.

Brownlee, I. A., Allen, A., Pearson, J. P., Dettmar, P. W., Havler, M. E., Atherton, M. R., & Onsoyen, E., (2005). Alginate as a source of dietary fiber. *Crit. Rev. Food. Sci. Nutr.*, *45*, 497–510.

Chakraborty, S., Khandai, M., Sharma, A., Khanam, N., Patra, C. N., Dinda, S. C., & Sen, K. K., (2010). Preparation, *in vitro* and *in vivo* evaluation of albino-pectinate bioadhesive microspheres: An investigation of the effects of polymers using multiple comparison analysis. *Acta Pharm.*, *60*, 255–266.

Chan, A. W., & Neufeld, R. J., (2010). Tuneable semi-synthetic network alginate for absorptive encapsulation and controlled release of protein therapeutics. *Biomater.*, *31*, 9040–9047.

Chan, L. W., Lee, H. Y., & Heng, P. W. S., (2006). Mechanisms of external and internal gelation and their impact on the functions of alginate as a coat and delivery system. *Carbohydr. Polym.*, *63*, 176–187.

Chang, D., Lei, J., Cui, H., Lu, N., Sun, Y., & Zhang, X., (2012). Disulfide cross-linked nanospheres from sodium alginate derivative for inflammatory bowel disease: Preparation, characterization, and *in vitro* drug release behavior. *Carbohydr. Polym.*, *88*, 663–669.

Chaudhuri, P. K., & Kar, M., (2005). Preparation of alginate gel beads containing metformin hydrochloride using the emulsion-gelation method. *Trop. J. Pharm. Res.*, *4*, 489–493.

Cheng, Y., Yu, S., Zhen, X, Wang, X., Wu, W., & Jiang, X., (2012). Alginic acid nanoparticles prepared through counterion complexation method as a drug delivery system. *ACS Appl. Mater. Interf.*, *4*, 5325–5332.

Ciofani, G., Raffa, V., Obata, Y., Menciassi, A., Dario, P., & Takeoka, S., (2008). Magnetic driven alginate nanoparticles for targeted drug delivery. *Curr. Nanosci.*, *4*, 212–218.

Cong-Gui, C., Gerelt, B., Shao, T. J., Nishiumi, T., & Suzuki, A., (2006). Effects of high pressure on pH, water-binding capacity and textural properties of pork muscle gels containing various levels of sodium alginate. *Asian-Australas. J. Anim. Sci.*, *19*, 1658–1664.

Dai, Y. N., Lin, P., Zhang, J. P., Wang, A. Q., & Wei, Q., (2008). Swelling characteristics and drug delivery properties of nifedipine-loaded pH-sensitive alginate-chitosan hydrogel beads. *J. Biomed. Mater. Res. Part B: Appl. Biomater.*, *86B*, 493–500.

Das, B., Dutta, S., Nayak, A. K., & Nanda, U., (2014). Zinc alginate-carboxymethyl cashew gum microbeads for prolonged drug release: Development and optimization. *Int. J. Biol. Macromol.*, *70*, 505–515.

Das, B., Nayak, A. K., & Nanda, U., (2013). Topical gels of lidocaine HCl using cashew gum and Carbopol 940: Preparation and *in vitro* skin permeation. *Int. J. Biol. Macromol.*, *62*, 514–517.

Das, M. K., & Senapati, P. C., (2008). Furosemide-loaded alginate microspheres prepared by ionic cross-linking technique: Morphology and release characteristics. *Indian J. Pharm. Sci.*, *70*, 77–84.

Das, R. K., Kasoju, N., & Bora, U., (2010). Encapsulation of curcumin in alginate-chitosan-pluronic composite nanoparticles for delivery to cancer cells. *Nanomed.*, *6*, 153–160.

DeRamos, C. M., Irwin, A. E., Nauss, J. L., & Stout, B. E., (1997). 13C NMR and molecular modeling studies of alginic acid binding with alkaline earth and lanthanide metal ions. *Inorg. Chimica Acta.*, *256*, 69–75.

Draget, K. I., Braek, G. S., & Smidsrod, O., (1994). Alginic acid gels: The effect of alginate chemical composition and molecular weight. *Carbohydr. Polym.*, *25*, 31–38.

Draget, K. I., Smidsrod, O., & Skjàk-Bræk, G., (2005). Alginate from algae. In: Steinbüchel, A., & Rhee, S. K., (eds.), *Polysaccharide and Polyamides in the Food Industry* (pp. pp. 1–30). Wiley-VCH Verlag GmbH & Co.: Weinheim.

Dusel, R., McGinity, J., Harris, M. R., Vadino, W. A., & Cooper, J., (1986). Sodium alginate. *Handbook of Pharmaceutical Excipients* (pp. 257–258). Published by American Pharmaceutical Society: the USA and the Pharmaceutical Society of Great Britain, London.

Efentakis, M., & Koutlis, A., (2001). Release of furosemide from multiple-unit and single-unit preparations containing different viscosity grades of sodium alginate. *Pharm. Dev. Tech., 6,* 91–98.

El-Kamel, A. H., Al-Gohary, O. M. N., & Hosny, E. A., (2003). Alginate-diltiazem hydrochloride beads: Optimization of formulation factors, *in vitro* and *in vivo* bioavailability. *J. Microencapsul., 20,* 211–225.

Ertesvag, H., & Valla, S., (1998). Biosynthesis and applications of alginates. *Polym. Degrad. Stabil., 59,* 85–91.

Fuoss, R. M., & Strauss, U. P., (1948). Electrostatic interaction of polyelectrolytes and simple electrolytes. *J. Polymer Sci., 3,* 602–603.

Gao, C. M., Liu, M. Z., Chen, S. L., Jin, S. P., & Chen, J., (2009). Preparation of oxidized sodium alginate-graftpoly ((2-dimethylamino) ethyl methacrylate) gel beads and *in vitro* controlled release behavior of BSA. *Int. J. Pharm., 371,* 16–24.

Gao, C., Tang, F., Zhang, J. X., Lee, S. M. Y., & Wang, R., (2017). Glutathione-responsive nanoparticles from a sodium alginate derivative for selective release of doxorubicin in tumor cells. *J. Mater. Chem. B., 5,* 2337–2346.

Gaumann, A., Laudes, M., Jacob, B., Pommersheim, R., Laue, C., Vogt, W., et al., (2000). Effect of media composition on long-term *in vitro* stability of barium alginate and polyacrylic acid multilayer microcapsules. *Biomater., 21,* 1911–1917.

George, M., & Abraham, T. E., (2006). Polyionic hydrocolloids for the intestinal delivery of protein drugs: Alginate and chitosan- a review. *J. Control. Release, 114,* 1–14.

Goh, C. H., Heng, P. W. S., & Chan, L. W., (2012). Alginates as a useful natural polymer for microencapsulation and therapeutic applications. *Carbohydr. Polym., 88,* 1–12.

Gombotz, W. R., & Wee, S. F., (1998). Protein release from alginate matrices, *Adv. Drug Deliv. Rev., 31,* 267–285.

Gonzalez-Rodriguez, M. L., Holgado, M. A., Sanchez-Lafuente, C., Rabasco, A. M., & Fini, A., (2002). Alginate/chitosan particulate systems for sodium diclofenac release. *Int. J. Pharm., 232,* 225–234.

Guru, P. R., Bera, H., Das, M., Hasnain, M. S., & Nayak, A. K., (2018). Aceclofenac-loaded *Plantago ovata*, F. husk mucilage-Zn^{+2}-pectinate controlled-release matrices. *Starch–Stärke, 70,* 1700136.

Guru, P. R., Nayak, A. K., & Sahu, R. K., (2013). Oil-entrapped sterculia gum-alginate buoyant systems of aceclofenac: Development and *in vitro* evaluation. *Colloids Surf. B: Biointerf., 104,* 268–275.

Haidar, Z. S., Hamdy, R. C., & Tabrizian, M., (2008). Protein release kinetics for core-shell hybrid nanoparticles based on the layer-by-layer assembly of alginate and chitosan on liposomes. *Biomater., 29,* 1207–1215.

Halder, A., Maiti, S., & Sa, B., (2005). Entrapment efficiency and release characteristics of polyethyleneimine-treated or -untreated calcium alginate beads loaded with propranolol-resign complex. *Int. J. Pharm., 302,* 84–94.

Haque, S., Mohammed, S., Sahni, J. K., Ali, J., & Baboota, S., (2014). Development and evaluation of brain-targeted intranasal alginate nanoparticles for the treatment of depression. *J. Psychiatr. Res., 48,* 1–12.

Hasnain, M. S., & Nayak, A. K., (2018). Alginate-inorganic composite particles as sustained drug delivery matrices. In: Inamuddin, A. A. M., & Mohammad, A., (eds.),

Applications of Nanocomposite Materials in Drug Delivery (pp. 39–74). A volume in Woodhead Publishing Series in Biomaterials, Elsevier Inc.

Hasnain, M. S., Nayak, A. K., Singh, M., Tabish, M., Ansari, M. T., & Ara, T. J., (2016). Alginate-based bipolymeric-nanobioceramic composite matrices for sustained drug release. *Int. J. Biol. Macromol.*, *83*, 71–77.

Hasnain, M. S., Nayak, A. K., Singh, R., & Ahmad, F., (2010). Emerging trends of natural-based polymeric systems for drug delivery in tissue engineering applications. *Sci. J. UBU.*, *1*, 1–13.

Hasnain, M. S., Rishishwar, P., Rishishwar, S., Ali, S., & Nayak, A. K., (2018a). Extraction and characterization of cashew tree (*Anacardium occidentale*) gum, use in aceclofenac dental pastes. *Int. J. Biol. Macromol.*, *116*, 1074–1081.

Hasnain, M. S., Rishishwar, P., Rishishwar, S., Ali, S., & Nayak, A. K., (2018b). Isolation and characterization of *Linum usitatisimum* polysaccharide to prepare mucoadhesive beads of diclofenac sodium. *Int. J. Biol. Macromol.*, *116*, 162–172.

Haug, A., Larsen, B., & Smidsrød, O., (1966). A study of the constitution of alginic acid by partial hydrolysis. *Acta Chem. Scand.*, *20*, 183–190.

Haug, A., Larsen, B., & Smidsrød, O., (1967). Studies on the sequence of uronic acid residues in alginic acid. *Acta Chem. Scand.*, *21*, 691–704.

Holte, Ø., Onsøyen, E., Myrvold, R., & Karlsen, J., (2003). Sustained release of water-soluble drug from directly compressed alginate tablets. *Eur. J. Pharm. Sci.*, *20*, 403–407.

Hua, S., Ma, H., Li, X., Yang, H., & Wang, A., (2010). pH-sensitive sodium alginate/poly(vinyl alcohol) hydrogel beads prepared by combined Ca^{2+} crosslinking and freeze-thawing cycles for controlled release diclofenac sodium. *Int. J. Biol. Macromol.*, *46*, 517–523.

Ishak, R. A. H., Awad, G. A. S., Mortada, N. D., & Nour, S. A. K., (2007). Preparation, *in vitro* and *in vivo* evaluation of stomach-specific metronidazole-loaded alginate beads as local anti-*Helicobacter pylori* therapy. *J. Control. Release*, *119*, 207–214.

Jain, S., & Amiji, M., (2012). Tuftsin-modified alginate nanoparticles as a noncondensing macrophage-targeted DNA delivery system. *Biomacromol.*, *13*, 1074–1085.

Jana, S., Das, A., Nayak, A. K., Sen, K. K., & Basu, S. K., (2013). Aceclofenac-loaded unsaturated esterified alginate/gellan gum microspheres: *In vitro* and *in vivo* assessment. *Int. J. Biol. Macromol.*, *57*, 129–137.

Jana, S., Gangopadhaya, A., Bhowmik, B. B., Nayak, A. K., & Mukhrjee, A., (2015b). Pharmacokinetic evaluation of testosterone-loaded nanocapsules in rats. *Int. J. Biol. Macromol.*, *72*, 28–30.

Jana, S., Manna, S., Nayak, A. K., Sen, K. K., & Basu, S. K., (2014). Carbopol gel containing chitosan-egg albumin nanoparticles for transdermal aceclofenac delivery. *Colloids Surf. B: Biointerf.*, *114*, 36–44.

Jana, S., Samanta, A., Nayak, A. K., Sen, K. K., & Basu, S. K., (2015a). Novel alginate hydrogel core-shell systems for combination delivery of ranitidine HCl and aceclofenac. *Int. J. Biol. Macromol.*, *74*, 85–92.

Javadzadeh, Y., Hamedeyazdan, S., Adibkia, K., Kiafar, F., Zarintan, H. M., & Barzegar-Jalali, M., (2010). Evaluation of drug release kinetics and physicochemical characteristics of metronidazole floating beads based on calcium silicate and gas-forming agents. *Pharm. Dev. Technol.*, *15*, 329–338.

Jayapal, J. J., & Dhanaraj, S., (2017). Exemestane loaded alginate nanoparticles for cancer treatment: Formulation and *in vitro* evaluation. *Int. J. Biol. Macromol., 105*, 416–421.

Jena, A. K., Nayak, A. K., De, A., Mitra, D., & Samanta, A., (2018). Development of lamivudine containing multiple emulsions stabilized by gum odina. *Future J. Pharm. Sci., 4*, 71–79.

Jimenez-Escrig, A., & Sanchez-Muniz, F. J., (2000). Dietary fiber from edible seaweeds. Chemical structure, physicochemical properties, and effects on cholesterol metabolism. *Nutri. Res., 20*, 585–598.

Jin, L. W., Chao, Q. W., Ren, X. Z., & Si, X. C., (2014). Multi-drug delivery system based on alginate/calcium carbonate hybrid nanoparticles for combination chemotherapy. *Colloids Surf. B: Biointerf., 123*, 498–505.

Khalid, M. Z., Fatima, Z., Mohammad, Z., Saima, R., & Mirza, N. A., (2015). Alginate-based polyurethanes: A review of recent advances and perspective. *Int. J. Biol. Macromol., 79*, 377–387.

Klaudianos, S., (1972). Slow-release alginate tablets. Effect of technological factors on the release of active material: Part 1. *Pharm. Ind., 34*, 976–982.

Kumar, S., Bhanjana, G., Verma, R. K., Dhingra, D., Dilbaghi, N., & Kim, K. H., (2017). Metformin-loaded alginate nanoparticles as an effective anti-diabetic agent for controlled drug release. *J. Pharm. Pharmacol., 69*, 143–150.

Li, L., Li, J., Si, S., Wang, L., Shi, C., Sun, Y., Liang, Z., & Mao, S., (2015). Effect of formulation variables on *in vitro* release of a water-soluble drug from chitosan-sodium alginate matrix tablets. *Asian J. Pharm. Sci., 10*, 314–321.

Liao, S. H., Liu, C. H., Bastakoti, B. P., Suzuki, N., Chang, Y., Yamauchi, Y., & Wu, K. C. W., (2015). Functionalized magnetic iron oxide/alginate core-shell nanoparticles for targeting hyperthermia. *Int. J. Nanomed., 10*, 3315–3328.

Liew, C. V., Chan, L. W., Ching, A. L., & Heng, P. W. S., (2006). Evaluation of sodium alginate as drug release modifier in matrix tablets. *Int. J. Pharm., 309*, 25–37.

Lin, Y. H., Liang, H. F., Chung, C. K., Chen, M. C., & Sung, H. W., (2005). Physical crosslinked alginate/N, O-carboxymethyl chitosan hydrogels with calcium for oral delivery of protein drugs. *Biomater., 26*, 2105–2113.

Ma, N., Xu, L., & Wang, Q. F., (2008). Development and evaluation of new sustained release floating microspheres. *Int. J. Pharm., 358*, 82–90.

Malakar, J., & Nayak, A. K., (2012a). Formulation and statistical optimization of multiple-unit ibuprofen-loaded buoyant system using 2^3-factorial design. *Chem. Eng. Res. Des., 9*, 1834–1846.

Malakar, J., & Nayak, A. K., (2012b). Theophylline release behavior for hard gelatin capsules containing various hydrophilic polymers. *J. Pharm. Educ. Res., 3*, 10–16.

Malakar, J., Das, K., & Nayak, A. K., (2014a). *In situ* cross-linked matrix tablets for sustained salbutamol sulfate release– Formulation development by statistical optimization. *Polim. Med., 44*, 221–230.

Malakar, J., Dutta, P., Purokayastha, S. D., Dey, S., & Nayak, A. K., (2014b). Floating capsules containing alginate-based beads of salbutamol sulfate: *In vitro-In vivo* evaluations. *Int. J. Biol. Macromol., 64*, 181–189.

Malakar, J., Nayak, A. K., & Das, A., (2013a). Modified starch (cationized)-alginate beads containing aceclofenac: Formulation optimization using central composite design. *Starch–Stärke, 65*, 603–612.

Malakar, J., Nayak, A. K., & Pal, D., (2012). Development of cloxacillin loaded multiple-unit alginate-based floating system by emulsion–gelation method. *Int. J. Biol. Macromol.*, *50*, 138–147.

Malakar, J., Nayak, A. K., Jana, P., & Pal, D., (2013b). Potato starch-blended alginate beads for prolonged release of tolbutamide: Development by statistical optimization and *in vitro* characterization. *Asian J. Pharm.*, *7*, 43–51.

Mandal, S., Basu, S. K., & Sa, B., (2009). Sustained release of a water-soluble drug from alginate matrix tablets prepared by wet granulation method. *AAPS Pharm. Sci. Tech.*, *10*, 1348–1356.

Mandal, S., Basu, S. K., & Sa, B., (2010). Ca^{2+} ion cross-linked interpenetrating network matrix tablets of polyacrylamide-grafted-sodium alginate and sodium alginate for sustained release of diltiazem hydrochloride. *Carbohydr. Polym.*, *82*, 867–873.

Martinez, A., Iglesias, I., Lozano, R., Teijon, J. M., & Blanco, M. D., (2011). Synthesis and characterization of thiolated alginate-albumin nanoparticles stabilized by disulfide bonds. Evaluation as drug delivery systems. *Carbohydr. Polym.*, *83*, 1311–1321.

Meng, X., Li, P., Wei, Q., & Zhang, H. X., (2011). pH-sensitive alginate-chitosan hydrogel beads for carvedilol delivery. *Pharm. Dev. Technol.*, *16*, 22–28.

Morshad, M. M., Mallick, J., Nath, A. K., Uddin, M. Z., Dutta, M., Hossain, M. A., & Kawsar, M. H., (2012). Effect of barium chloride as a cross-linking agent on the sodium alginate based diclofenac sodium beads. *Bangladesh Pharm. J.*, *15*, 53–57.

Nahar, K., Hossain, M. K., & Khan, T. A., (2017). Alginate and its versatile application in drug delivery. *J. Pharm. Sci. Res.*, *9*, 606–617.

Nayak, A. K., & Pal, D., (2011). Development of pH-sensitive tamarind seed polysaccharide-alginate composite beads for controlled diclofenac sodium delivery using response surface methodology. *Int. J. Biol. Macromol.*, *49*, 784–793.

Nayak, A. K., & Pal, D., (2012). Natural polysaccharides for drug delivery in tissue engineering. *Everyman's Sci.*, *XLVI*, 347–352.

Nayak, A. K., & Pal, D., (2013a). Ionotropically-gelled mucoadhesive beads for oral metformin HCl delivery: Formulation, optimization and antidiabetic evaluation. *J. Sci. Ind. Res.*, *72*, 15–22.

Nayak, A. K., & Pal, D., (2013b). Formulation optimization of jackfruit seed starch-alginate mucoadhesive beads of metformin HCl. *Int. J. Biol. Macromol.*, *59*, 264–272.

Nayak, A. K., & Pal, D., (2014). *Trigonella foenum-graecum* L. seed mucilage-gellan mucoadhesive beads for controlled release of metformin HCl. *Carbohydr. Polym.*, *107*, 31–40.

Nayak, A. K., & Pal, D., (2016). Plant-derived polymers: Ionically gelled sustained drug release systems, In: Mishra, M., (ed.), *Encyclopedia of Biomedical Polymers and Polymeric Biomaterials* (Vol. 8, pp. 6002–6017). Taylor & Francis Group, New York, NY 10017, U.S.A.

Nayak, A. K., & Pal, D., (2017a). Natural starches-blended ionotropically-gelled microparticles/beads for sustained drug release, In: Thakur, V. K., Thakur, M. K., & Kessler, M. R., (eds.), *Handbook of Composites from Renewable Materials, Nanocomposites: Advanced Applications* (Vol. 8, pp. 527–560). Wiley-Scrivener, USA.

Nayak, A. K., & Pal, D., (2017b). Tamarind seed polysaccharide: An emerging excipient for pharmaceutical use. *Indian J. Pharm. Educ. Res.*, *51*, S136–S146.

Nayak, A. K., & Pal, D., (2018). Functionalization of tamarind gum for drug delivery, In: Thakur, V. K., & Thakur, M. K., (eds.), *Functional Biopolymers* (pp. 35–56). Springer International Publishing, Switzerland.

Nayak, A. K., (2010). Advances in therapeutic protein production and delivery. *Int. J. Pharm. Pharmaceut. Sci.*, *2*, 1–5.

Nayak, A. K., (2016). Tamarind seed polysaccharide-based multiple-unit systems for sustained drug release, In: Kalia, S., & Averous, L., (eds.), *Biodegradable and Bio-Based Polymers: Environmental and Biomedical Applications* (pp. 471–494). Wiley-Scrivener, USA.

Nayak, A. K., Ahmad, S. A., Beg, S., Ara, T. J., & Hasnain, M. S., (2018a). Drug delivery: Present, past and future of medicine, In: Inamuddin, A. A. M., & Mohammad, A., (eds.), *Applications of Nanocomposite Materials in Drug Delivery* (pp. 255–282). A volume in Woodhead Publishing Series in Biomaterials, Elsevier Inc.

Nayak, A. K., Ara, T. J., Hasnain, M. S., & Hoda, N., (2018b). Okra gum-alginate composites for controlled releasing drug delivery, In: Inamuddin, A. A. M., & Mohammad, A., (eds.), *Applications of Nanocomposite Materials in Drug Delivery* (pp. 761–785). A volume in Woodhead Publishing Series in Biomaterials, Elsevier Inc.

Nayak, A. K., Beg, S., Hasnain, M. S., Malakar, J., & Pal, D., (2018). Soluble starch-blended Ca^{2+}-Zn^{2+}-alginate composites-based microparticles of aceclofenac: Formulation development and *in vitro* characterization. *Future J. Pharm. Sci.*, *4*, 63–70.

Nayak, A. K., Bera, H., Hasnain, M. S., & Pal, D., (2018c). Graft-copolymerization of plant polysaccharides, In: Thakur, V. K., (ed.), *Biopolymer Grafting, Synthesis and Properties* (pp. 1–62). Elsevier Inc.

Nayak, A. K., Das, B., & Maji, R., (2012b). Calcium alginate/gum Arabic beads containing glibenclamide: Development and *in vitro* characterization. *Int. J. Biol. Macromol.*, *51*, 1070–1078.

Nayak, A. K., Hasnain, M. S., & Pal, D., (2018d). Gelled microparticles/beads of sterculia gum and tamarind gum for sustained drug release, In: Thakur, V. K., & Thakur, M. K., (eds.), *Handbook of Springer on Polymeric Gel* (pp. 361–414). Springer International Publishing, Switzerland.

Nayak, A. K., Hasnain, M. S., Beg, S., & Alam, M. I., (2010). Mucoadhesive beads of gliclazide: Design, development and evaluation. *Sci. Asia.*, *36*, 319–325.

Nayak, A. K., Pal, D., & Hasnain, M. S., (2013b). Development and optimization of jackfruit seed starch-alginate beads containing pioglitazone. *Curr. Drug Deliv.*, *10*, 608–619.

Nayak, A. K., Pal, D., & Malakar, J., (2013c). Development, optimization and evaluation of emulsion-gelled floating beads using natural polysaccharide-blend for controlled drug release. *Polym. Eng. Sci.*, *53*, 338–350.

Nayak, A. K., Pal, D., & Santra, K., (2013d). Plantago *ovata* F. Mucilage-alginate mucoadhesive beads for controlled release of glibenclamide: Development, optimization, and *in vitro-in vivo* evaluation. *J. Pharm*. Article ID 151035.

Nayak, A. K., Pal, D., & Santra, K., (2015). Screening of polysaccharides from tamarind, fenugreek and jackfruit seeds as pharmaceutical excipients. *Int. J. Biol. Macromol.*, *79*, 756–760.

Nayak, A. K., Pal, D., & Santra, K., (2016). Swelling and drug release behavior of metformin HCl-loaded tamarind seed polysaccharide-alginate beads. *Int. J. Biol. Macromol.*, *82*, 1023–1027.

Nayak, A. K., Pal, D., Pany, D. R., & Mohanty, B., (2010). Evaluation of *Spinacia oleracea* L. leaves mucilage as innovative suspending agent. *J. Adv. Pharm. Technol. Res.*, *1*, 338–341.

Nayak, A. K., Pal, D., Pradhan, J., & Ghorai, T., (2012). The potential of *Trigonella foenum-graecum* L. seed mucilage as suspending agent. *Indian J. Pharm. Educ. Res.*, *46*, 312–317.

Nayak, A. K., Pal, D., Pradhan, J., & Hasnain, M. S., (2013a). Fenugreek seed mucilage-alginate mucoadhesive beads of metformin HCl: Design, optimization and evaluation. *Int. J. Biol. Macromol.*, *54*, 144–154.

Nesamony, J., Singh, P. R., Nada, S. E., Shah, Z. A., & Kolling, W. M., (2012). Calcium alginate nanoparticles synthesized through a novel interfacial cross-linking method as a potential protein drug delivery system. *J. Pharm. Sci.*, *101*, 2177–2184.

Ojantakanen, S., Hannula, A. M., Marvola, M., Klinge, E., & Makioaa, M., (1993). Bioavailability of ibuprofen from hard gelatin capsules containing sucralfate or sodium alginate. *Eur. J. Pharm. Biopharm.*, *39*, 197–201.

Pal, D., & Nayak, A. K., (2011). Development, optimization and anti-diabetic activity of gliclazide-loaded alginate-methyl cellulose mucoadhesive microcapsules. *AAPS Pharm. Sci. Tech.*, *12*, 1431–1441.

Pal, D., & Nayak, A. K., (2012). Novel tamarind seed polysaccharide-alginate mucoadhesive microspheres for oral gliclazide delivery. *Drug Deliv.*, *19*, 123–131.

Pal, D., & Nayak, A. K., (2015a). Alginates, blends and microspheres: Controlled drug delivery, In: Mishra, M., (ed.), *Encyclopedia of Biomedical Polymers and Polymeric Biomaterials* (Vol. 1, pp. 89–98). Taylor & Francis Group, New York, NY 10017, U.S.A.

Pal, D., & Nayak, A. K., (2015b). Interpenetrating polymer networks (IPNs): Natural polymeric blends for drug delivery. In: Mishra, M., (ed.), *Encyclopedia of Biomedical Polymers and Polymeric Biomaterials* (Vol. 5, pp. 4120–4130). Taylor & Francis Group, New York, NY 10017, U.S.A.

Pal, D., & Nayak, A. K., (2017). Plant polysaccharides-blended ionotropically-gelled alginate multiple-unit systems for sustained drug release. In: Thakur, V. K., Thakur, M. K., & Kessler, M. R., (eds.), *Handbook of Composites from Renewable Materials, Polymeric Composites* (Vol. 6, pp. 399–400). Wiley-Scrivener, U.S.A.

Papageorgiou, M., Kasapis, S., & Gothard, M. G., (1994). Structural and textural properties of calcium-induced, hot-made alginate gels. *Carbohydr. Polym.*, *24*, 199–207.

Paraskevopoulou, A., Boskou, D., & Kiosseoglou, V., (2005). Stabilization of olive oil–lemon juice emulsion with polysaccharides. *Food Chem.*, *90*, 627–634.

Park, K. A., (1989). New approach to study mucoadhesion: Colloidal gold staining. *Int. J. Pharm.*, *53*, 209–217.

Pasparakis, G., & Bouropoulos, N., (2006). Swelling studies and *in vitro* release of verapamil from calcium alginate-chitosan beads. *Int. J. Pharm.*, *323*, 34–42.

Patel, Y. L., Sher, P., & Pawar, A. P., (2006). The effect of drug concentration and curing time on processing and properties of calcium alginate beads containing metronidazole by response surface methodology. *AAPS Pharm. Sci. Tech.*, *7*, Article 86.

Pawar, S. N., & Edgar, K. J., (2011). Chemical modification of alginates in organic solvent systems. *Biomacromol.*, *12*, 95–103.

Pawar, S. N., & Edgar, K. J., (2012). Alginate derivatization: A review of chemistry, properties and applications. *Biomater.*, *33*, 3279–3305.

Polk, A., Amsden, B., Yao, K. D., Peng, T., & Goosen, M. F. A., (1994). Controlled release of albumin from chitosan-alginate microcapsules. *J. Pharm. Sci.*, *83*, 178–185.

Pornsak, S., Nartaya, T., & Kingkarn, K., (2007). Swelling, erosion and release behavior of alginate-based matrix tablets. *Eur. J. Pharm. Biopharm.*, *66*, 435–450.

Potter, K., & McFarland, E. W., (1996). Ion transport studies in calcium alginate gels by magnetic resonance microscopy. *Solid State Nucl. Mag. Reson.*, *6*, 323–331.

Racovita, S., Vasilu, S., Popa, M., et al., (2009). Polysaccharides based micro- and nanoparticles obtained by ionic gelation and their applications as drug delivery systems. *Rev. Roma. de Chim.*, *54*, 709–718.

Rees, D. A., & Welsh, E. J., (1977). Secondary and tertiary structure of polysaccharides in solutions and gels. *Angewandte Chem.*, *16*, 214–224.

Rees, D. A., (1981). Polysaccharide shapes and their interactions. Some recent advances. *Pure Appl. Chem.*, *53*, 1–14.

Rehm, B. H. A., & Valla, S., (1997). Bacterial alginates: Biosynthesis and applications. *Appl. Microbiol. Biotechnol.*, *48*, 281–288.

Santhi, K., Dhanraj, S. A., Nagasamyvenkatesh, D., Sangeetha, S., & Suresh, B., (2005). Preparation and optimization of sodium alginate nanospheres of methotrexate, *Indian J. Pharm. Sci.*, *67*, 691–696.

Saralkar, P. L., & Dash, A. K., (2017). Alginate nanoparticles containing curcumin and resveratrol: Preparation, characterization, and *in vitro* evaluation against DU145 prostate cancer cell. *AAPS Pharm. Sci. Tech.*, *18*, 2814.

Sherys, A. Y., Gurov, A. N., & Tolstoguzov, V. B., (1989). Water-insoluble triple complexes: Bovine serum albumin–bivalent metal cation–alginate. *Carbohydr. Polym.*, *10*, 87–102.

Shilpa, A., Agrawal, S. S., & Ray, A. R., (2003). Controlled delivery of drugs from alginate matrix. *J. Macromol. Sci. Part C. Polym. Rev.*, *43*, 187–221.

Singh, B., Sharma, V., & Chauhan, D., (2010). Gastroretentive floating sterculia-alginate beads for use in antiulcer drug delivery. *Chem. Eng. Res. Des.*, *88*, 997–1012.

Singh, I., Kumar, P., Singh, H., Goyal, M., & Rana, V., (2011). Formulation and evaluation of domperidone loaded mineral oil entrapped emulsion gel (MOEG) buoyant beads. *Acta Pol. Pharm. - Drug Res.*, *68*, 121–126.

Sinha, P., Ubaidulla, U., & Nayak, A. K., (2015a). Okra (*Hibiscus esculentus*) gum-alginate blend mucoadhesive beads for controlled glibenclamide release. *Int. J. Biol. Macromol.*, *72*, 1069–1075.

Sinha, P., Ubaidulla, U., Hasnain, M. S., Nayak, A. K., & Rama, B., (2015b). Alginate-okra gum blend beads of diclofenac sodium from aqueous template using ZnSO$_4$ as a cross-linker. *Int. J. Biol. Macromol.*, *79*, 555–563.

Sirkia, T., Salonen, H., Veski, P., Jurjenson, H., & Marvola, M., (1994). Biopharmaceutical evaluation of new prolonged-release press-coated ibuprofen tablets containing sodium alginate to adjust drug release. *Int. J. Pharm.*, *107*, 179–187.

Skjaak-Braek, G., Grasdalen, H., & Larsen, B., (1986). Monomer sequence and acetylation pattern in some bacterial alginates. *Carbohydr. Res.*, *154*, 239–250.

Smidsroed, O., Glover, R. M., & Whittington, S. G., (1973). Relative extension of alginates having different chemical composition. *Carbohydr. Res.*, *27*, 107–118.

Smrdel, P., Bogataj, M., & Mrhar, A., (2008). The influence of selected parameters on the size and shape of alginate beads prepared by ionotropic gelation. *Sci. Pharm.*, *76*, 77–89.

Sriamornsak, P., & Sungthongjeen, S., (2007). Modification of theophylline release with alginate gel formed in hard capsules. *AAPS Pharm. Sci. Tech.*, *8*, Article 51.

Sriamornsak, P., Thirawong, N., & Korkerd, K., (2007). Swelling, erosion and release behavior of alginate-based matrix tablets. *Eur. J. Pharm. Biopharm.*, *66*, 435–450.

Stockwell, A. F., Davis, S. S., & Walker, S. E., (1986). *In vitro* evaluation of alginate gel systems as sustained release drug delivery systems. *J. Control. Release*, *3*, 167–175.

Streubel, A., Siepmann, J., & Bodmeier, R., (2002). Floating microparticles based on low-density foam powder. *Int. J. Pharm.*, *241*, 279–292.

Sutherland, L. W., (1991). Alginates. In: Byron, D., (ed.), *Biomaterials: Novel Materials from Biological Sources* (pp. 307–332). New York: Stockton.

Takka, S., & Acartürk, F., (1999). Calcium alginate microparticles for oral administration: I-Effect of sodium alginate type on drug release and drug entrapment efficiency. *J. Microencapsul.*, *16*, 275–290.

The United States Pharmacopoeia, Sodium Alginate, (2000). (24th edn., p. 2515). United States Pharmacopoeial Convention, Inc., Rockville, MD.

Tonnesen, H. H., & Karlsen, J., (2002). Alginate in drug delivery systems. *Drug Dev. Ind. Pharm.*, *28*, 621–630.

Torre, M. L., Giunchedi, P., Maggi, L., Stefli, R., Mechiste, E. O., & Conte, U., (1998). Formulation and characterization of calcium alginate beads containing ampicillin. *Pharm. Dev. Technol.*, *3*, 193–198.

Vandenberg, G., Drolet, C., Scott, S., & De La Noüe, J., (2001). Factors affecting protein release from alginate-chitosan coacervate microcapsules during production and gastric/intestinal simulation. *J. Control. Release*, *77*, 297–307.

Verma, A., Dubey, J., Verma, N., & Nayak, A. K., (2017). Chitosan-hydroxypropyl methylcellulose matrices as carriers for hydrodynamically balanced capsules of moxifloxacin HCl. *Curr. Drug Deliv.*, *14*, 83–90.

Veski, P., & Marvola, M., (1993). Sodium alginate as diluents in hard gelatine capsule containing ibuprofen as a model drug. *Pharmazie*, *48*, 757–760.

Veski, P., Marvola, M., Smal, J., Heiskanen, I., & Jurjenson, H., (1994). Biopharmaceutical evaluation of pseudoephedrine hydrochloride capsules containing different grades of sodium alginate. *Int. J. Pharm.*, *111*, 171–179.

Wang, X., Wenk, E., Hu, X., Castro, G. R., Meinel, L., Wang, X., Li, C., Merkle, H., & Kaplan, D. L., (2007). Silk coatings on PLGA and alginate microspheres for protein delivery. *Biomater.*, *28*, 4161–4169.

Wells, L. A., & Sheardown, H., (2007). Extended release of high pI proteins from alginate microspheres via a novel encapsulation technique. *Eur. J. Pharm. Biopharm.*, *65*, 329–335.

Wittaya-areekul, S., Kruenate, J., & Prahsarn, C., (2006). Preparation and *in vitro* evaluation of mucoadhesive properties of alginate/chitosan microparticles containing prednisolone. *Int. J. Pharm.*, *312*, 113–118.

Yang, J. S., Xie, Y. J., & He, W., (2011). Research progress on chemical modification of alginate: A review. *Carbohydr. Polym.*, *84*, 33–39.

Yao, H., Yao, H., Zhu, J., Yu, J., & Zhang, L., (2012). Preparation and evaluation of a novel gastric floating alginate/poloxamer inner-porous beads using foam solution. *Int. J. Pharm.*, *422*, 211–219.

Yegin, B. A., Moulari, B., Durlu-Kandilci, N. T., Korkusuz, P., Pellequer, Y., & Lamprecht, A., (2007). Sulindac loaded alginate beads for a mucoprotective and controlled drug release. *J. Microencapsul.*, *24*, 371–382.

Zhang, C., Wang, W., Liu, T., Wu, Y., Guo, H., Wang, P., Tian, Q., Wang, Y., & Yuan, Z., (2012). Doxorubicin-loaded glycyrrhetinic acid-modified alginate nanoparticles for liver tumor chemotherapy. *Biomater.*, *33*, 2187–2196.

Zhao, D., Zhuo, R. X., & Cheng, S. X., (2012). Alginate modified nanostructured calcium carbonate with enhanced delivery efficiency for gene and drug delivery. *Mol. Biosyst.*, *8*, 753–759.

Zheng, H., (1997). Interaction mechanism in sol-gel transition of alginate solutions by addition of divalent cations. *Carbohydr. Res.*, *302*, 97–101.

CHAPTER 3

Pharmaceutical Applications of Agar-Agar

MD SHAHRUZZAMAN, SHANTA BISWAS, MD NURUS SAKIB,
PAPIA HAQUE, MOHAMMED MIZANUR RAHMAN, and
ABUL K. MALLIK

Department of Applied Chemistry and Chemical Engineering, Faculty of Engineering and Technology, University of Dhaka, Dhaka 1000, Bangladesh

ABSTRACT

Over the last decade, much interest has been developed in producing polymer-based materials from natural sources owing to their biocompatibility, biodegradability, affordability, availability, and nontoxic behavior. Among the enormous variety of natural polymers, Agar-Agar is a unique natural polymer that increasingly preferred over synthetic materials in addition to being considered alternative sources of raw materials for pharmaceutical applications. Due to its outstanding intrinsic properties, for example, the strength of the gel produced from Agar-Agar, it is highly sought in the pharmaceutical industry. The pharmaceutical applications of Agar-Agar are mostly as an agent for gelation, stabilization, and thickening. In addition to that, purgative use of Agar-agar and as surgical aid is also common. Many works have been done by researchers to obtain Agar-based materials such as a composite hydrogel, nanocomposite film, etc. to use in pharmaceutical sectors. An injectable and phase-changeable composite hydrogel made from Agar was prepared for treating tumors having the use in both chemo and photothermal therapy. This composite hydrogel was able to load and subsequently release chemotherapeutics and antibiotics. Moreover, Agar-based nanocomposite film showed efficacy in inhibiting the growth of *Listeria monocytogenes*. Blends of Agar and different polysaccharides are also gaining importance in pharmaceutical aspects. This chapter gives an overview of the key features of inherent properties of Agar-Agar, their modification, with special attention to potential pharmaceutical applications.

3.1 INTRODUCTION

In recent years, polymeric materials sourced from nature are on the spotlight in the fields of pharmaceutical research. Natural polymers are thought to have an edge on their synthetic counterparts because they are more copious, biocompatible, and biodegradable (Vendruscolo et al., 2009). Among natural polymers, Agar-Agar is a unique natural polymer that increasingly preferred over synthetic materials in addition to being considered alternative sources of raw materials for pharmaceutical applications. It has found its use as a digestive regulator, surgical aid, and in drug delivery. Agar-Agar is also used as a food ingredient; mainly as a gelling agent, for example, in making jellies, candies, and dessert items (Figure 3.1). Agar-Agar is recognized to be a safe product by the United States Food and Drug Administration Ordinance (USFDA) and also meet the requirements of the food chemicals codex (FCC).

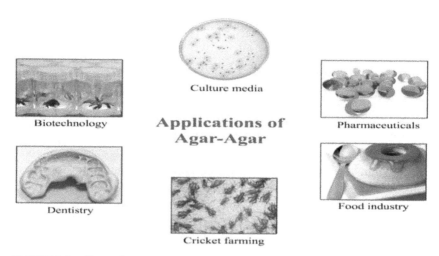

FIGURE 3.1   (See color insert.) Potential applications of agar-agar.

Agar-Agar is a jelly-like substance, processed from certain red algae of the *Rhodophyta phylum*. It is a mixture of polysaccharides, which include galactose, 3,6-anhydrogalactose and inorganic sulfate bonded to the carbohydrate. Agar-Agar is composed of two components. It has the agarose; a linear polysaccharide, and blend of smaller molecules, which are called agaropectin (Figure 3.2) (Williams and Glyn, 2000). Its solubility in water increases with higher temperature.

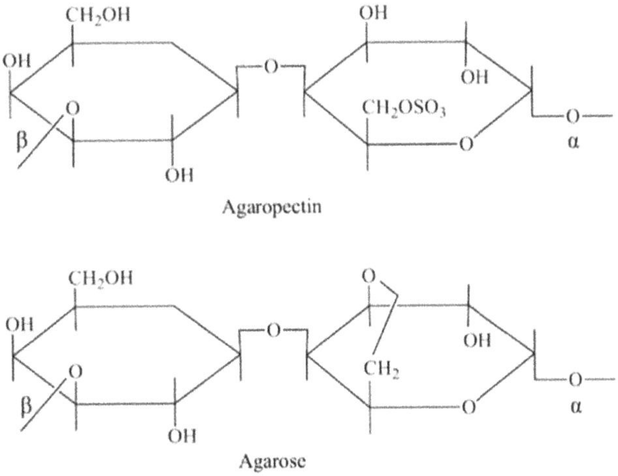

FIGURE 3.2   Structures of agarose and agaropectin.

Agar-Agar, the phycocolloid was first discovered in Japan by Minoya Tarozaemon in the year 1658. At that time, the use of Agar-agar was mostly in the east. Eventually, the application of Agar-agar has spread, and currently, it is used in almost all around the world. Originally Agar-agar was known by the term 'tokoroten' and was used extensively near factory areas. It was sold in both hot, extracted as a solution, and cold, in the gel form. It was then industrialized as a dry product named 'kanten.' The word 'Agar-Agar' is taken of Malayan origin and is commonly used. It is also known as 'gelosa' mainly in French- and Portuguese-speaking countries. The extent of application of Agar-agar is increasing owing to the high gel strength, high solubility, and excellent transparency, which are unlike of other phycocolloid, gelatin, or gum. Hence, Agar-Agar has been in extensive use and sometimes exclusively used in research as well as industries.

Several types of Agar are present among which, pharmaceutical agar, blood agar, sabouraud agar, chocolate agar, nutrient, and non-nutrient agar, tryptic soy agar, neomycin agar, etc. are most common. Pharmaceutical agar, as the name specifies, is utilized in the pharmaceutical industry for laxatives, excipient ingredient in pills, tablets, and syrups, preparation of medicaments, and so on. This chapter gives an overview of the key features of inherent properties of Agar-Agar, their modification, with special attention to potential applications in the pharmaceutical industry.

3.2   SOURCES OF AGAR-AGAR

The Agar-Agar is simply called as Agar. Chemically, Agar is a polymer composed of sugar units Galactose. It was first introduced in the west at the end of the nineteenth century. It was the first phycocolloid used by man as well as the very first foodstuff approved by FDA. The first agar was developed in Japan from *Gelidium amansii*. This 'red' (Rhodo-phyceae phyllum) seaweed type was available in large quantities along their respective coasts. As the other type (*Gelidium amansii)* was scarce, *Rhodophyceaes* was used as substitutes (Armisén and Galatas, 2009).

Gracilaria seaweeds give an agar of very poor gelling property. So, it was transformed into a stronger final product by some chemical transformation. The process will be discussed in the following section. A natural internal transformation, also known as maturing of polysaccharide in *Gelidium, Gelidiella, and Pterocladia* seaweeds, is carried out by an enzymatic process (Armisén and Galatas, 2009). In the following Figure 3.3, different species that are used for producing agar are given.

3.3   METHODS OF EXTRACTION OF AGAR-AGAR

Agar-Agar is produced in many variants, viz. flakes, powder, thread, and bars. Usually, powdered agar-agar is used for industrial applications. Bar, thread, and flake types are used for cooking purpose. A general technique for separating a polysaccharide from Seaweed involves a combination of extraction, cooling, Filtration, Drying, etc. At first, a definite weight of weed sample is soaked in a sodium phosphate buffer. After that, it is pressurized in a pressure bomb apparatus starting from a temperature of 110°C to end at 140°C. Next, Cooling at 90–95°C and pressure filtration removes all weed debris giving a fresh extract. However, the extract is re-extracted by a buffer, which ultimately undergoes filtration, free-thawing, and drying operation, giving a polysaccharide (Falshaw et al., 1998). Finally, it is milled and screened to have the desired Agar product.

Phylum: Rodophyta.
Class: Florideophyceae.
    Order: Gelidiales.
        Family: Gelidiaceae.
            Genus: *Gelidium.*
                Species: *G. sesquipedale*,*
                        G. amansii*,
                        G. robustum*,
                        G. pristoides,
                        G. canariense.
                        G. rex,
                        G. chilense.
                Genus: *Gelidiella.*
                  Species: *G. acerosa.*
                Genus: *Pterocladia.*
                  Species: *P. capillacea*,*
                        P. lucida*
    Order: Gracilariales.
        Family: Gracilariaceae.
            Genus: *Gracilaria.*
                  Species: *G. chilense*,*
                        G. gigas,
                        G. edulis,
                        G. gracilis,
                        G. tenuistipitata
                Genus: Gracilariopsis
                  Species: *G. lamaneiformis.*
                        G. sjostedtii
    Order: Ahnfeltiales.
        Family: Ahnfeltiaceae.
            Genus: *Ahnfeltia.*
                Species: *A. plycata.*

* The most used agarophytes in industry

FIGURE 3.3 **(See color insert.)** Different species of the agarophytes. Reprinted with permission. Copyright reserved by Elsevier (Armisén and Galatas, 2009).

However, there are three different techniques to extract Agar from the seaweed, which gives Agar product of different quality. They are:

- Traditional method of extracting agar;
- Freezing-thawing method; and
- Syneresis method.

i) **Traditional Method:** In the middle of the seventeenth century, Minoya developed a technique of producing 'natural Agar' in shapes of square agar (Kaku-Kanten) or strip agar (Ito-Kanten). Making of such types requires the careful cleaning of the seaweed at the primary step. *Gelidium amansii* species is usually preferred for the first step (washing) by employing similar devices that were used to clean tealeaves. After that, the seaweed was carefully picked and stored so that no contamination was done. Then the desired product was extracted by means of pH-adjusted hot water.

In earlier days, sake or vinegar was used to adjust pH. Presently diluted sulfuric acid is used. The extract was then sifted or filtered by passing it through cotton bags. The filtered extract was poured in wooden trays and cooled. Then, it was cut to either square-shaped bars ($4 \times 6 \times 24$ cm^3) or to strips of 25–40 cm. Finally, it was placed onto bamboo grills and kept for freezing all night. After one or two nights when the agar is completely frozen then water is sprayed over to thaw. After that, the product was sun-dried and stored safely. This ancient technique yielded uneven shapes and low-quality Agar, and this was unsuitable for production owing to the heavy dependence in climate (Armisén and Galatas, 2009).

ii) **Freezing-Thawing Method:** "Natural Agar" was first produced by this traditional method until 1939 when American Agar & Co. (San Diego, USA) started producing Agar on an industrial scale. It is based on the technique of making the ice bars in the freezing tanks. Only 1% and 1.2% of Agar is present in the seaweed extract. After thawing and straining by centrifugation produces a final product, which contains 10% to 12% of Agar. The effluent water contains salts, oligomers, and as well as proteins phycoerythrins that are responsible for the color of the algae. In this process, water is not removed completely, and it might contain some water-soluble impurities, which gives an Agar of poor quality (Armisén and Galatas, 2009).

iii) **Syneresis Method:** Syneresis method applies the gelling colloid property that enables absorbed water to be removed by applying appropriate force. In Japan, this technique was applied uniquely for *Gracilaria* Agar. This extract is then filtered using the cloth as the filter medium primarily against the stone block and pressed by means of small hydraulic presses for removing residual water (Armisén and Galatas, 2009).

The method was spread throughout the World by building new plants in many countries. At first, this process was semi-automated and later modified to a completely automated one. It is also available in the form of a vertical mechanical press. Around the world, this method is more followed as it is more economical. This is because the former method requires an additional ice requirement of 80/100 metric tons for each ton of agar. Although both of the methods can be mechanized today, syneresis has reduced energy cost. Agar purity is high in case of Syneresis method because most of the water had been removed and with this water, most of the soluble impurities are also removed, and syneresis agars have lower ash content (Figure 3.4). However, many customers do not care about this better quality Agar (Armisén and Galatas, 2009).

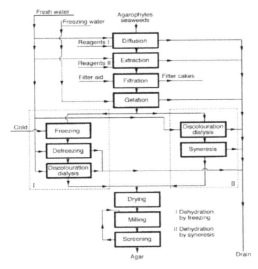

FIGURE 3.4 Agar fabrication diagram.
Reprinted with permission, Copyright reserved by Elsevier (Armisén and Galatas, 2009).

Sometimes, alkali pretreatment is mandatory prior to the extraction of polysaccharide from seaweed. This technique involves many chemicals including NaOH/NaBH$_4$ (4%/0.4% w/v), NaH$_2$PO$_4$ (7.5% w/v), sodium phosphate buffer, etc. under the operation of soaking, decantation, washing, etc. After this pretreatment, the regular procedure of separating polysaccharide is followed as described in the previous paragraph (Falshaw et al., 1998).

3.4 PROPERTIES OF AGAR-AGAR

Cell walls of agarophytes algae contain agar as the structural carbohy-drate. Generally, it exists in the as calcium salt or a blend of calcium and magnesium salts. Agar is composed of two polysaccharides named as: (i) Agarose, which is a neutral polymer; and (ii) Agaropectin, which is a sulfated charged polymer. Agarose is the component mainly responsible for the gelling action of the Agar. Agarose is sulfate free linear neutrally charged molecule which contains chains of repeating alternate units of β-1,3-linked- D-galactose and α-1,4-linked 3,6-anhydro-L-galactose. Whereas, Agaropectin, is the sulfated-non-gelling fraction (3% to 10% sulfate) with varying percentages of sulfate ester, D-glucuronic acid, and pyruvic acid. More than half of the naturally found agar-agar is agarose. However, the ratio varies on the basis of the seaweed species.

Agarose gels are considered as physical gels because of the presence of bonds between hydrogen and agarose molecules, firmed by structural water. The mechanism involved in gelation is governed completely by hydrogen bonding. These gels contain large pores of water with a thick Agarose chains and exhibits as a high turbidity and elasticity. The network can be formed even 0.1% w/w Agarose concentrations. It is also believed that Agarose gels are formed due to the ordered formation of sol particles upon cooling from a heated solution. Helices are formed as cooling progress, which allows the helics to aggregate forming a gel network (Barrangou et al., 2006). In the case of organic solvents, Agarose is normally insoluble and cannot form gels. Apparently, the structure of water highly determines the gelling possibility in aqueous solutions (Fernández et al., 2008).

Beside a physical gel, it is also considered as a thermo-reversible gel. Agarose portion is a double helical format, which aggregates to make a 3-D framework for holding the water molecules. Thus, a thermo-reversible gel of Agar is formed (Armisén and Galatas, 2009). The mechanical behavior of the aqueous Agar-Agar gel is also quite interesting (Fernández et al., 2008). The gelling portion shows thermal hysteresis, which ultimately forms large aggregates. This aggregate can withstand temperatures much higher than individual helices (Fernández et al., 2008).

In terms of the chemistry of the gelling in Agar, it is necessary to shift the sulfate group from 6-O-sulfo-L-galactopyranosyl residues (an enzyme-mediated process) forming a closed ring structure. It is believed that in the Golgi apparatus chains of alternating D- and L-galactopyranosyl residues are aggregate on primer molecules. At the initial stage, sulfation of L-galac-topyranosyl residues thought to occur in the Golgi, whereas the ring closure and methylation is believed to happen at the later stage (Hemmingson et al., 1996). Methyl ether substituents are known to have an impact on the properties of agar gels. When methylation is occurred naturally, it can alter both the gel setting and melting temperatures. But if the methylation is synthetic, it is found to have lowered the gel-setting and gel-melting temperature (Falshaw et al., 1998). Agars from *C. codioides, Curdiea angustata, C. crassa* and *C. abellata* were methylated on position 6 of the 3-linked β-D-galactopyranosyl units, and agars originating from *C. racovitzae, C. irwinii* were almost fully methylated on position 2 of the 4-linked 3,6-anhydro-α-L-galactopyranosyl units. However, the Agars from *C. coriacea* and *C. obese* were almost completely methylated at both these positions (Falshaw et al., 1998). Initially, when agars from *Gracilaria* seaweeds were extracted, they had low gelation and termed as agaroids. After the discovery of alkaline hydrolysis of sulfates by Yanagawa, Agars from *Gracilaria* produced stronger gel strength (Armisén and Galatas, 2009).

Agar solution viscosity varies depending on the source of raw materials. At pH 4.5 to 9.0 and above gelling temperature, the viscosity is almost constant. At pH range 6.0 to 8.0, the solution is not influenced by ionic strength or age. However, viscosity increases by time at constant temperature once gelling has been started. In case of a charge, an Agar-Agar solution is slightly negative. Electric charge and hydration property largely dominate its stability. If the two are absent, Agar-Agar solution may flocculate. If the Agar-Agar is exposed to a higher temperature for a long time, the gel strength is found to be low. This process happens faster if the pH is lowered. Therefore, Agar-Agar solutions should not be at high temperature and low pH (not lower than 6.0) for longer times. If dried, Agar-Agar do not contaminate by microorganisms. However, gels and solutions of Agar-Agar are considered as a nutrient media for microorganism, and hence, precautions are required to inhibit the growth. Some properties of extracted Agar-Agar are listed in Table 3.1.

TABLE 3.1   Properties of Extracted Agar-Agar

Property	Extracted Agar-Agar
Moisture content (%)	6.8 ± 0.3
Ash (%)	3.8 ± 3.1
pH	8.5 ± 0.0
Gel strength (at 25°C)	608.3 ± 30.1
Gelling temperature	37.0 ± 2.0
Melting temperature	76.0 ± 1.5
Viscosity	3.9 ± 0.1
Sulfate (%)	2.0 ± 0.2
Pyruvate (%)	8.8 ± 1.1
3,6-Anhydrogalactose (%)	24.4 ± 2.7

3.5 APPLICATIONS OF AGAR-AGAR

3.5.1 DIGESTIVE REGULATOR

Agar-Agar is high on fiber, so it has the potential to improve the digestion by promoting the movement of food in digestive tracts (Armisén and Galatas, 2009). Once ingested, agar triples in size by absorbing water (12–15 times its weight of fluids) making a fuller sensation in the stomach. Agar-Agar E406 helps in removing toxic substances of the intestine and decreases blood sugar level.

Since Agar-Agar takes in glucose and leaves the digestive system fast, it does not allow the body to retain excess fats. It also absorbs water, which eventually helps in removing body waste. In addition, Agar-Agar takes up bile and hence help the body to process more cholesterol.

3.5.2 SURGICAL AND MEDICAMENT PREPARATION

Agar-Agar has diversified applications in the pharmaceutical industries. Agar-agar is used for preparing capsules, incipient in tablets, aiding surgical action by providing lubrication and acting as the suspending agent in radiology (Figure 3.5). Sulfated agar has the ability to lower lipid levels in blood like heparin. The most common use of Agar-Agar in the laboratory is probably in the culture medium for the bacterium.

3.5.3 OTHER APPLICATIONS

Apart from pharmaceutical use, Agar-Agar has diverse applications in many fields such as an ingredient in the culture medium, pesticide detector, and especially in food preparation.

Use of silver nanoparticles in Agar matrix has found its application as an antimicrobial thin film. In the process, very stable silver nanoparticles are mixed with Agar matrix to prepare a thin organic-inorganic hybrid film. This film can be reused; however, due to the loss of silver nanoparticles, the efficiency drops. The Antimicrobial test results show that it is most effective for *C. albicans* and retained its microbial property even after three cycles of reuse. Antimicrobial films may be promising for food packaging, wound dressing, and as sterile coatings (Ghosh et al., 2010).

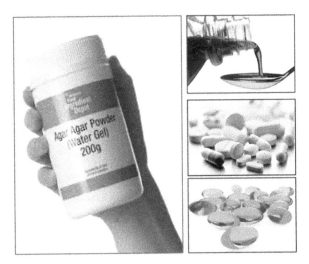

FIGURE 3.5 **(See color insert.)** Pharmaceutical applications of agar-agar.

In agrochemical delivery, Agar-based super sorbent, pH sensitive, the hydrogel has been prepared by Pourjavadi et al., The maximum water absorbency in distilled water was more than 1000 g/g. Potassium nitrate was loaded to check the efficiency of the hydrogel for the sustained release of potassium. Copolymerization of acrylic acid and 2-acrylamido-2-methylpropane sulfonic acid on agar produced a hydrogel, which released potassium sharply in the initial state and leveled off afterward (Pourjavadi et al., 2009).

Food packaging films can also be made from Agars. A study shows that *Gracilaria vermiculophylla* extracted agar with glycerol as a plasticizer showed potential in better oxygen permeability as edible coatings. Use of plasticizer had increased the elongation characteristics (Sousa et al., 2010).

Agar gel is often used as a culture medium. A novel method was applied to create colonies of human bone marrow cells using agar gel as a medium. White blood cells were used as a stimulus for colonial growth. The garulocytic colonies reached 500–1500 cells at 12–16 days (Pike and Robinson, 1970).

Another study on agar shows that it lowers cholesterol absorption. Kelley and Tsai have conducted an experiment on rats to find the effects of complex carbohydrates on cholesterol absorption. They used agar, pectin, and gum Arabic as carbohydrates. The study showed that all three lowered the cholesterol absorption (Kelley and Tsai, 1978).

3.6   PHARMACEUTICAL APPLICATIONS OF BLENDS OF AGAR-AGAR

Blending of polymers allows them to be tailored to specific properties of interest. Physical and Mechanical properties of natural gums can be further enhanced by blending them with other polymers. Biocompatibility, biodegradation, release rate controlling, etc. properties are mainly modified by blending. Due to their synergistic effect, the blending of polymers help overcome their limitations and can show good potential in pharmaceutical applications (Laurienzo, 2010).

Blends of agar and naturally occurring polysaccharides or agar and synthetic polymers are of interest owing to the fact that such blends are able to form reversible gels by changing the temperature of the solutions. Blending with agar generally enhances the water holding capacity and improved gelation properties (Laurienzo, 2010).

In drug delivery, the improvement of natural gums by blending can be significant. Generally, single polymers tend to release the drugs faster, especially when the drugs are hydrophilic in nature. Thus blending helps in achieving the sustainable and prolonged release of drugs. Agar-gelatin blends have been investigated for the delivery of salbutamol drug as a tablet format. Salbutamol helps to inhibit hypersensitive mast cells and to relax the bronchial muscles. Generally, salbutamol is administered through the use of inhalers for direct and efficient administration as well as for the best results. However, tablet format, using agar-gelatin blends, can be used to administer salbutamol to persons who are unable to use inhalers. The blends were a mixture of agar and two types of gelatin; Gelatin A and B. Gelatin A is processed by means of acid and Gelatin B is processed by means of alkali. Different blends of agar and gelatin were put under the test of drug release, and the results showed that the blends showed better results in acting as release controlling agents than components alone. Blend of agar and gelatin A & B held the drug the longest. The suggested reasons attributed to the higher controlling characteristics were the interactions between the groups Gel A and Gel B and hydrogen bond formation. The release kinetics showed a quick release initially followed by a moderate one. Such types are important when immediate relief is required (Saxena et al., 2011).

Another agar and gelatin blends have been studied to view the Theophylline drug delivery. In the process, three blended hydrogels were prepared, and their drug release behaviors were studied. The three blends were agar and gelatin, gelatin, and κ–carrageenan, and agar and κ-carrageenan hydrogels. Of the three, the agar and gelatin blend hydrogel showed the slowest release (Liu et al., 2005). Agar and polyvinyl alcohol (PVA) have been blended to produce hydrogel. In the process, agar, and PVA were mixed physically in different ratios and observed their characteristics such as swelling ratio, mechanical characteristics, and thermal characteristics. PVA hydrogels have a drawback of poor mechanical characteristics. The blended hydrogel showed higher melting temperature and improved mechanical characteristics. Furthermore, due to biocompatibility, these hydrogels may find use in biomedical applications such as wound healing (Lyons et al., 2009).

Agar and κ–carrageenan can be grafted with polyvinyl pyrrolidone (PVP) to make hydrogel. Agar, κ-carrageenan, and PVP blend hydrogel showed good water absorption capacity and crystallinity. Hence they can be a potential medium for targeted applications, in biomedicine and in pharmaceuticals as hydrogel dressings (Prasad et al., 2006).

Another use for Agar and κ-carrageenan hydrogel blend was put to use as a carrier for enzyme immobilization. κ-carrageenan and agar gel were reacted with both polyethyleneimine and glutaraldehyde (GA) to give the specific substituents for binding enzymes. β-D-galactosidase (β-gal) enzyme remained around 97.7% after its fifteenth reusability cycle (Figure 3.6). This gel disk was found to be superior to treated agar gel and less expensive than the commercial counterparts (Wahba and Hassan, 2017).

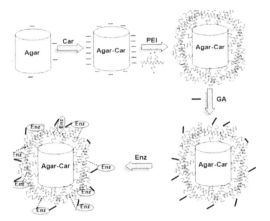

FIGURE 3.6   (See color insert.) Illustration showing steps in preparing and activating agar-car gel disk for immobilization of enzymes.
Reprinted with permission, Copyright reserved by Springer (Wahba and Hassan, 2017).

Hydrogel prepared from grafting agar and sodium alginate blend with acrylamide showed superior absorbency and pH resistance. Polyacrylamide was grafted to agar and sodium alginate blend. It is acidic and alkaline stable with low shear thinning behavior (Meena et al., 2008). Hydrogel made using agar, and sodium alginate blend with acrylonitrile showed similar characteristics. In this process, agar, and sodium alginate blend was graft copolymerized with acrylonitrile using potassium persulphate as an initiator. Study shows that the hydrogel was stable over a pH range of 1.2 to 12.5. Due to its stable nature, it may find its use in newer applications (Chhatbar et al., 2009).

In another study, blends of food hydrocolloids, agar, κ-carrageenan were cross-linked with genipin to make hydrogel. The hydrogel was highly viscous and thermally stable. Along with the swelling ability, the weight loss ratio was low (Meena et al., 2009).

3.7   FUTURE SCOPE OF AGAR-AGAR

The present world is shifting towards natural sources. Agar and their blends are promising materials for pharmaceutical industries as well as for other uses. Components made from natural materials such as agar are generally more biodegradable and bio-compatible than synthetic materials. Presently, blending natural materials with synthetic ones are common and can make a great deal of improvements in the shortcomings of natural materials. For example, agar, and alginate-based hydrogel beads, responsive to change in pH has shown potential in controlled release oral drug delivery. The mechanical strength has also been improved due to blending compared to native alginate beads (Yin et al., 2018). Silk fiber and agar composites showed better biocompatibility, and the composite was not rejected from the body (Thu-Hien et al., 2018). Furthermore, since agar can be amply produced in the nature, this source of polysaccharide can be termed as a renewable and sustainable source. Agar has all the potential to contribute significantly to the modern world's pharmaceutical needs.

3.8   CONCLUDING REMARKS

This chapter presented the structure and properties of Agar-Agar and the pharmaceutical applications of Agar-Agar and Agar-based materials. Due to the biocompatibility, biodegradability, low price and availability, Agar-Agar is a very attractive and potential candidate for pharmaceutical applications, especially as a digestive regulator, surgical lubricants, in the preparation of medicaments. Another important field of applications of Agar-based materials is as pharmaceutical excipients, which have been continuing to lead the research efforts of scientists to find cheaper, biodegradable, and eco-friendly excipients. Hence it can be said that Agar-Agar has some clear benefits over their synthetic counterparts due to their intrinsic properties, and this can provide a window of opportunities for its applications.

KEYWORDS

- **Agar-Agar**
- **biodegradability**
- **blends**
- **composite**
- **hydrogel**
- **pharmaceutical**

REFERENCES

Armisén, R., & Galatas, F., (2009). Agar. *Handbook of Hydrocolloids* (2nd edn., pp. 82–107). Woodhead Publishing.

Barrangou, L. M., Daubert, C. R., & Foegeding, E. A., (2006). Textural properties of agarose gels. I. Rheological and fracture properties. *Food Hydrocoll., 20*(2/3), 184–195.

Chhatbar, M. U., et al., (2009). *Agar/Sodium Alginate-Graft-Polyacrylonitrile, a Stable Hydrogel System*, 1085–1090.

Falshaw, R., Furneaux, R. H., & Stevenson, D. E., (1998). Agars from nine species of red seaweed in the genus curdiea (Gracilariaceae, Rhodophyta). *Carbohydr. Res., 308*(1/2), 107–115.

Fernández, E., et al., (2008). Rheological and thermal properties of agarose aqueous solutions and hydrogels. *J. Polym. Sci. B Polym. Phys., 46*(3), 322–328.

Ghosh, S., et al., (2010). Antimicrobial activity of highly stable silver nanoparticles embedded in agar-agar matrix as a thin film. *Carbohydr. Res., 345*(15), 2220–2227.

Hemmingson, J. A., Furneaux, R. H., & Murray-Brown, V. H., (1996). Biosynthesis of agar polysaccharides in Gracilaria chilensis Bird, McLachlan et Oliveira. *Carbohydr. Res., 287*(1), 101–115.

Kelley, J. J., & Tsai, A. C., (1978). Effect of pectin, gum Arabic and agar on cholesterol absorption, synthesis, and turnover in rats. *J. Nutr., 108*(4), 630–639.

Laurienzo, P., (2010). Marine polysaccharides in pharmaceutical applications: An overview. *Mar. Drugs, 8*(9), 2435.

Liu, J., et al., (2005). Release of theophylline from polymer blend hydrogels. *Int. J. Pharm., 298*(1), 117–125.

Lyons, J. G., et al., (2009). Development and characterization of an agar–polyvinyl alcohol blend hydrogel. *J. Mech. Behav. Biomed. Mater., 2*(5), 485–493.

Meena, R., et al., (2008). Development of a robust hydrogel system based on agar and sodium alginate blend. *Polym. Int., 57*(2), 329–336.

Meena, R., Prasad, K., & Siddhanta, A., (2009). Development of a stable hydrogel network based on agar–kappa-carrageenan blend cross-linked with genipin. *Food Hydrocoll., 23*(2), 497–509.

Pike, B. L., & Robinson, W. A., (1970). Human bone marrow colony growth in agar-gel. *J. Cell. Physiol., 76*(1), 77–84.

Pourjavadi, A., Farhadpour, B., & Seidi, F., (2009). Synthesis and investigation of swelling behavior of new agar based superabsorbent hydrogel as a candidate for agrochemical delivery. *J. Polym. Res., 16*(6), 655.

Prasad, K., et al., (2006). Hydrogel-forming agar-graft-PVP and κ-carrageenan-graft-PVP blends: Rapid synthesis and characterization. *J. Appl. Polym. Sci., 102*(4), 3654–3663.

Saxena, A., et al., (2011). Effect of agar—gelatin compositions on the release of salbutamol tablets. *Int. J. Pharma. Investig., 1*(2), 93–98.

Sousa, A. M., et al., (2010). In: *Biodegradable Agar Extracted from Gracilaria vermiculophylla: Film Properties and Application to Edible Coating* (pp. 739–744). Materials science forum, Trans Tech Publ.

Thu-Hien, L., et al., (2018). Evaluation of the morphology and biocompatibility of natural silk fibers/agar blend scaffolds for tissue regeneration. *International J. Polym. Sci., 2018*, 7.

Vendruscolo, C. W., et al., (2009). Physicochemical and mechanical characterization of galactomannan from *Mimosa scabrella*: Effect of drying method. *Carbohydr. Polym., 76*(1), 86–93.

Wahba, M. I., & Hassan, M. E., (2017). Agar-carrageenan hydrogel blend as a carrier for the covalent immobilization of β-D-galactosidase. *Macromol. Res., 25*(9), 913–923.

Williams, P., & Glyn, P., (2000). *Handbook of Hydrocolloids*. Cambridge: Woodhead, ISBN 1-85573-501-6.

Yin, Z. C., Wang, Y. L., & Wang, K., (2018). A pH-responsive composite hydrogel beads based on agar and alginate for oral drug delivery. *J. Drug Deliv. Sci. Technol., 43*, 12–18.

CHAPTER 4

Pharmaceutical Applications of Gellan Gum

SHAKTI NAGPAL, SUNIL KUMAR DUBEY, VAMSHI KRISHNA RAPALLI, and GAUTAM SINGHVI

Department of Pharmacy, Birla Institute of Technology & Science (BITS), Pilani, Rajasthan, 333031, India

ABSTRACT

Extensive studies were carried out over the past few decades in the area of microbial polysaccharides due to the beneficial properties. Evolving the opportunities in the pharmaceutical industry, the cosmeceutical industry, and nutraceutical industry have been possible because of the diversity in structural features. Native to Sphingomonas genus and being a sphingan hetro-polysaccharide, gellan is obtained by fermentation only. Amongst the group of linear bio-polysaccharides, gellan has been studied extensively. Repeating units of D-glucose, D-glucuronate, and L-Rhamnose forms the tetra-saccharide. This ability of gellan to act as a polyfunctional ingredient in the pharmaceutical industry as capsule shells, swellable matrix, gelling, and thickening agent, bio-adhesive, binding agent, disintegrant has led to diverse applications. The properties of gelation for gellan have been studied critically to develop the modified drug release formulations. Improvement in properties of formulations, including oral, ophthalmic, nasal has been studied. Applications have also been extended in the areas of regenerative medicine, biotechnology, somatology, and tissue engineering. Gellan gum is under exploration for its biomimetic properties, especially for soft tissue engineering. This chapter gives an overview of the vast applications of gellan gum.

4.1 INTRODUCTION

Natural gums have been an area of immense opportunity because of its exclusive physicochemical properties, lower production cost, biocompatible nature, easy availability, and non-toxicity (Osmalek et al., 2014). Apart from this advantageous nature, certain drawbacks are also associated, including microbial infestation, hysterical hydration, erratic viscosity drops on storage (Vijan et al., 2012). Thus, chemical modifications in its structure and hybridization are required to fabricate the properties of these natural polymers. Most compounds of this class consist of polysaccharides as seeds (konjac gum, guar gum (GG)); exudates (gum tragacanth, gum ghatti, gum acacia) from plants; microbial fermentation (xanthan gum (XG), welan gum, gellan gum etc.) and from seaweeds (carrageenan, alginates, agar-agar) (Osmalek et al., 2014).

Majorly these have been used throughout the world as a food additive. Also, an increased advent of natural products as an integral part of the healthy lifestyle is there. Reduction of the dietary fat component and increased fibers in diet, with an increased consciousness towards health benefits, is the need of today. On interacting with water, these polysaccharides gain newer texture, anticipate specific functions (stabilizer, fat replacer, water absorber, thickener) (Prajapati et al., 2013).

Among various polysaccharide based natural polymer, gellan gum has been investigated for various pharmaceutical and biomedical application. Gellan gum is a bacterial polysaccharide which is obtained from *Sphingomonas elodea*. It was identified to possess a commercial advantage in 1978 during an extensive soil-water borne bacterial screening program by C.P. Kelco (USA). Produced from the strains of *Sphigomonas*, Gellan gum consists of rod-shaped, gram-negative, strict aerobe, chemoheterotrophic bacteria which produce yellowish pigmented colonies and consists of glycosphingolipids in their cell wall. Gellan is a linear polymer, anionic in nature and has a repeating sequence of tetra-saccharides consisting of two residues of β-D-Glucose (L-Rhamnose and β-D-Glucuronate) (Figure 4.1). The polysaccharide, which is biosynthesized has L-glyceryl substituents on O (2) and an acetyl group at O (6) of glucose unit in tetra-saccharide repeat sequences. During its commercial production, both the substituents get removed when treated with hot alkali and fermentation broth (Prajapati et al., 2013).

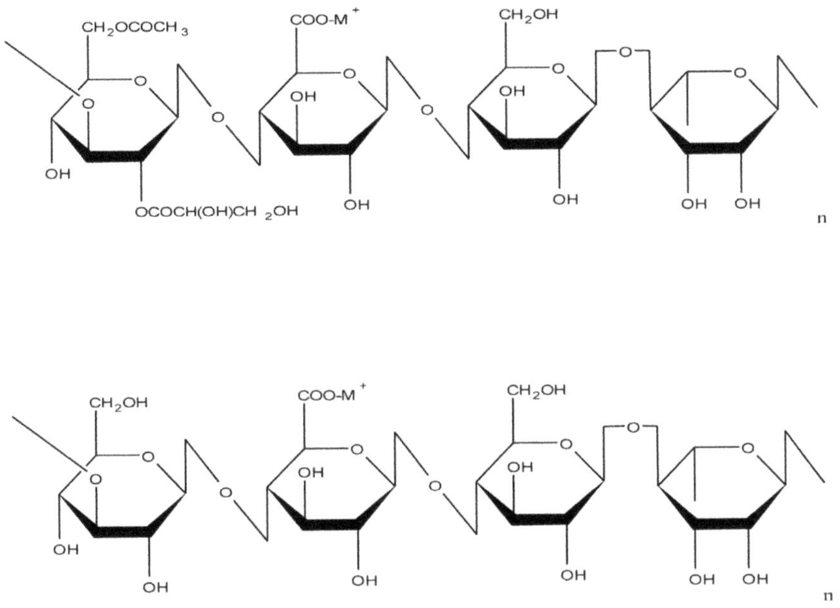

FIGURE 4.1 Chemical structure of (A) high acylated tetrasaccharide, (B) low acylated tetrasaccharide.

Gellan was approved by the US-FDA in November 1992 for being used as a source of food and then as an additive in food products in many other countries including South America, Canada, South Africa, Japan, etc. In 1996, at the 46th Joint Expert Committee on Food Additives, the specifications for gellan gum were established as shown in Table 4.1 (Osmalek et al., 2014).

The polysaccharide of plant and algae shows variation due to their molecular weight, biological source, conditions for plant growth and extraction methods of polysaccharides, etc. Though, it is expected to show less variability due to strict control over process parameters, including pH, aeration, temperature, release from fluid broth and, agitation. Thus, differences in the experimental procedure may occur. The counter ions like divalent cations, including calcium ion, magnesium ion, and monovalent cations, including sodium ion, potassium ion are used in balancing the charge of the carboxylate group in the polymer chain. Among these, the divalent cations are found to be more effective in enhancing the gelation properties of gellan as compared to monovalent cations.

The gellan gum is composed of repeating units of d-glucose (60%), l-rhamnose (20%) and d-glucuronic acid (20%) as well as some amounts of proteins and ashes, which can be removed by techniques such as filtration or centrifugation.

TABLE 4.1   General Specifications of Pharmaceutical Grade Gellan Gum (Prajapati et al., 2013)

Property	Value
Molecular weight	Near to 5,00,000
Appearance	Powdered form, off white, poor flowing
Application	Thickener, stabilizer, gel-forming agent
Solubility	Water soluble, insoluble in ethyl alcohol
Loss on drying	<15%
Nitrogen	<3%
Heavy metals (Lead)	Not more than 2 mg/kg
Microbial Contamination (*E. coli*, *Salmonella*, Yeast)	Negative for *E. coli*, *Salmonella*, and not more than 400 colonies/g of yeast.

4.2   CHEMISTRY OF GELLAN GUM

Gellan gum which is anionic in nature exists in two forms: one of which is the acetylated form usually known as high acyl gellan gum (HAGG) and other is deacylated form known as low acyl gellan gum (LAGG). The HAGG consists of two acyl groups, L-glycerate, O-acetate bound to its glucose unit. LAGG is obtained from HAGG by its alkaline hydrolysis. Both the acetylated and deacylated forms after the transition, coil to form a double helix structure and on cooling it forms (Morris et al., 2012a; Prajapati et al., 2013).

Gelation properties of gellan gum have been investigated extensively. The divalent cations (Ca^{2+} and Mg^{2+}) possess more efficient gelation properties as compared to the monovalent cations (Na^+ and K^+). In case of monovalent cations, the gelation is because of the repulsion through electrostatic forces developing on ionization of carboxylate groups in gellan polymer chains whereas gelation occurs as a result of chemical bonding between carboxyl groups of the glucuronic acid molecule and the gellan gum chain in case of divalent cations (Norton et al., 2011).

4.3 STRUCTURAL CHARACTERS OF GELLAN IN THE SOLID STATE

The X-ray diffraction of gellan gum has clearly depicted a double helix with each strand being 3-fold, aligned in levo fashion exhibiting a pitch of 5.64 nm. Parallel chains in half staggered confirmations, so that pitch is half that of the original strand, i.e., 2.82 nm. This arrangement is similar to that of carrageenan. The polysaccharide unit has four glycosidic linkages with a repeat sequence of monosaccharide units. The bonding is di-equatorial (1→4), and this leads to a flat and ribbon-like structure (Morris et al., 2012a).

4.3.1 CATIONIC INTERACTION

The presence of anionic groups in the polymer chain finds an attachment to poly-cations by electrostatic interaction. The strength of attachment is dependent on the linear charge density. Measurement of activity coefficient, osmotic pressure change, conductivity, and transport coefficient can be seen as cation activity (Morris et al., 2012a).

4.3.2 CONFORMATIONAL TRANSITIONS IN SOLUTION FORM

To study this, the strands have to convert to linear, to avoid any complications associated with different cations. Tetra-methyl ammonium salt helps to stabilize the strands. Optical rotation is a well-established technique to study the conformational transitions. An increase in temperature has parallelly been seen with the transition from an ordered structure at lower temperatures to a highly disordered structure at higher temperatures in the solution form (Morris et al., 2012a).

4.3.3 EFFECT OF EXCESS SALT AND LOW PH

An initial rise in gelation is attributed as a result of transiently increased salt concentration and later decrease in strength on further increasing salt concentrations. Reduction in pH has been ascribed as the best possible factor to induce gelation. This is not generally monotropic, but still, hydrochloric acid is observed to be the best agent for studying pH-induced changes in gelation. Similar to the salt and transient, a decrease in pH is effective for higher gelation. This effect can be seen with a pH change from neutral to 3.5. Further, reduction in pH is mostly followed best in isolation and purification of gellan during biosynthesis and fermentation (Morris et al., 2012a).

4.3.4   DEGRADATION BEHAVIOR

In ionic cross-linked hydrogels of gellan gum, a loss in stability and structure deterioration may occur *in-vivo* resulting in an exchange of divalent cations for monovalent cations. This could be overcome by combining with the inorganic materials or photoinitiated crosslinking by introducing reactive double bonds. Other methodologies adopted may include blending with biomolecules (Vieira et al., 2015).

4.4   METHACRYLATION OF GELLAN GUM

The methacrylation of gellan gum can be performed with chemically modifying LAGG. The photo-crosslinking of methacrylated gellan gum, causing an increase in the extent of cross-linking. The photo-crosslinked methacrylated gellan gum (phGG-MA) and ionic-crosslinked methacrylated gellan gum (iGG-MA) have been used for nucleus pulposus regeneration as alternative implants. The methacrylation of gellan gum by light irradiation results in the formation of hydrogels. Light irradiation initiates a free radical polymerization reaction and forms covalent cross-links between the polymer chains which leads to improved mechanical properties. For example, the value of Young's modulus as a result of cross-linking is between 0.15 and 148 kPa and those of gellan gum hydrogel having a value of 56 kPa which further affects degradation rate and *in-vitro* swelling kinetics (Vieira et al., 2015; Morris et al., 2012).

4.4.1   CHEMICAL SCISSORING (OXIDATION)

With the help of an oxidant like sodium periodate, the gellan gum chains are cleaved, leading to increase in the cross-linking potential of aldehyde groups and decrease the molecular weight of the polymer. Experimental observations by Tang et al. (2012), declare that the cleavage of cross-linking points due to oxidation can cause an increase in water absorption capacity. Through a Schiff base reaction, the cross-linkage amongst gellan and carboxymethyl chitosan (CHC-GG) enables gel structure stabilization. *In-vitro* studies showed that gellan hydrogels possess ability as such to be used in place of injectable vehicles for delivery of chondrocytes and due to its stabilization through Schiff base reaction, there is a reduction in the number of free aldehyde groups; resulting in enhanced chondrocyte proliferation (Vieira et al., 2015; Morris et al., 2012; Bacelar et al., 2016).

4.4.2 CARBOXYMETHYLATION AND THIOLATION

Carboxymethylation of gellan gum with the help of chloroacetic acid leads to the improvement of its aqueous solubility. For thiolation, gellan gum is incorporated with thiol groups containing L-cysteine using 1-ethyl-3-(3-dimethylaminopropyl) carbodiimide (EDC) activator, or it could be achieved by carrying out esterification of gellan with thioglycolic acid which could further increase the mucoadhesive properties and sensitivity towards cation- induced gelation. The carboxylation and thiolation of gellan gum improve its gelation in-situ by immobilizing the thiol and carboxymethyl groups (Morris et al., 2012; Bacelar et al., 2016).

4.4.3 GELATION TEMPERATURE

In physiological cationic conditions, the gelation temperature of unmodified gellan gum was found to be $> 42°C$ which is too high for injectable purposes thus the gelation temperature could be altered by oxidative cleavage which adjusts the molecular weight of gellan or by carboxymethylation and thiolation (Bacelar et al., 2016). The chemical modifications overcome the limitation of reduced physical stability and flexibility, reduced temperature spectrum for viable encapsulation of cell.

4.5 PRODUCTION OF GELLAN BY FERMENTATION

The media requirement may be specific for different exopolysaccharides by microbes. There may be a difference in the rate of biosynthesis and yield with a change in the composition of growth media. The media components include majorly sources of carbon, nitrogen, and inorganic salts. Abundant amounts of these may be required along with some necessary adjuvants like vitamins and special precursors as micronutrients, which may be added to promote the cellular growth and synthesis as a part of it. Critical factors affecting the production of gellan by fermentation have been tabulated (Table 4.2).

Isolation followed by purification has been an important part of successful fermentation. For isolation, the broth is first heated for 10–15 min at 90–95°C for cellular lysis, and viscosity reduction, and promoting mixing while precipitation. Centrifugation and filtration promote the separation of cellular debris. Precipitation is promoted by adding ice-cold isopropanol and by keeping the mixture at 4°C for 12 h. Precipitation is then followed by centrifugation and drying. The drying of the product is done at 55°C for 1 h. Clarification by filtration using 0.22μ filter and precipitation by isopropyl alcohol. Purification is carried out by subsequent treatment using organic solvents like acetone and ether. Also, it can be dissolved and dialyzed in a mass cut-off of 12,000–14,000 using dialysis cartridges. Further, gel filtration chromatography may also be carried out to purify the gum (Prajapati et al., 2013; Martin et al., 2014).

4.6   COMMERCIAL FORMS OF GELLAN GUM

a. **Gelrite:** A linear polymer comprised of various saccharides including glucose, glucuronic acid, O-acetyl moieties, and rhamnose. It is freely dispersible in water, either hot or cold, forming a viscous solution. This form can be incorporated to induce high gelation strength at lower concentrations of GELRITE, which can be as low as half that required for the agar. This form can withstand autoclaving conditions required for sterilization. Low viscosity liquid forms are seen when this form is subjected to higher temperatures. It is also resistant to degradation by enzymes. Being chemically inert, its additive must be heated above the gel point to assist integration (El-Laithy et al., 2011; Prajapati et al., 2013).

b. **Kelcogel:** Hydrocolloid material obtained from *Sphingomonas elodea*, by fermentation using direct carbohydrates. A high-acetylated product is obtained, which is then processed by alkali to make it deacetylated. It is available as a free flowing white powder,

TABLE 4.2 Factors Affecting the Fermentation of Gellan

SL. No.	Factor		Effect
1	Media Components	Carbon	Production yields (varies from 7.9–14.5g/L on varying sugars), composition, structure, and properties of the bacterial polysaccharide.
		Nitrogen	The gellan broth characteristics are strongly affected by the choice of nitrogen source being used. It was demonstrated by Daniel, Dreveton, Lonen in 1994 that the organic nitrogen was found to enhance the cell growth and biosynthesis of gellan gum.
		Precursors	Before the repeating unit of gellan gum is assembled, the gellan gum synthesis requires the presence of activated precursor. Bajaj et al., carried out various studies and demonstrated that at the concentration of 0.05%, tryptophan gave a maximum yield of gellan gum, i.e., 139.5g/L.
2	pH		pH value has been found to influence the formation of product and cell growth. The pH range desirable for the production of gellan varies from 6.5 to 7. It has been found that under acidic conditions, the production of gellan gum is significantly reduced. The optimum pH value for bacterial polysaccharide production has been found to be higher as compared to that of fungal glucan production.
3	Agitation Rate		For the proper mixing of gellan gum broth, a helical ribbon impeller was found appropriate to be used with an agitation rate of 250 rpm. Due to increased revolutions per minute, cavitation is increased and has also lead to the formation of stagnant layers in the gellan gum broth.
4	Dissolved oxygen and oxygen transfer capacity (DOT)		In a reduction in cellular growth, thus, synthesis is observed with an increased oxygen depletion. During the growth phase of gellan polymer, a high rate of production is marked, and it is also seen that DOT levels >20% do not affect the cell growth. The DOT levels increased up to 100% results in increased uptake rate of oxygen by cells and yield of gellan gum show a significant increase up to 23 g/L. A high level of DOT also leads to an increase in molecular mass and viscosity of the polymer.
5	Temperature		Maximum production is at 20–25°C, mostly carried out at 30°C but reduced above it.

and two forms are available, i.e., low acyl product and high acyl product.

c. **Gel-Gro:** It is a purified polysaccharide and obtained from *Pseudomonas elodea*. Chemically, rhamnose, glucose, glucuronic acid, and O-acetyl moieties are present in *Gel-Gro*. It appears as a dry white powder and contains up to 95% of solids. It produces high gel strength up to 225–500 g/cm^2 when media is formulated at 0.6–0.8% with 0.10% hydrated magnesium sulfate salts. *Gel-Gro* can be used as a substitute in place of agar. It has the capacity to form a clear gelatinous complex in the vicinity of cations. Similarly to agar in thermal stability, rigidity, compatibility with nutrient components, and clarity, this can be synonymously applied in microbiology as an ideal matrix to act as a nutrient media. Certain advantages compared to agar can be seen as,

- It can give similar qualities at just one-fourth to one-half the concentration of agar;
- Appearance is clearer;
- These possess higher stability to temperature as compared to agar;
- No contaminating material is present as compared to some toxic metabolites which often are seen with agar;
- Has also overcome the limitations of uncontrolled natural conditions affecting agar.

4.7 PHARMACEUTICAL APPLICATIONS OF GELLAN GUM

4.7.1 APPLICATIONS IN ORAL DELIVERY

The active component is intended to be delivered at a desired site in the body to exhibit its therapeutic activity. For a patient, depending on the compliance and convenience, the oral route is the most preferred one and therefore is studied extensively. The drug release from a formulation intended for oral route and comprising of a natural polymer depends on its ability of the polymer to swell, erode or diffuse in an aqueous environment. As a part of the complex interaction, the hydration of these gums from peripheral regions to core further limits the entry of fluid inside. As a result of it, the process of diffusion is hindered. The pharmaceutical applications have been summarized in Table 4.3.

TABLE 4.3 Formulations, Applications of Different Drugs Using Gellan Gum

Drug	Formulation	Action	Application	Reference
Clarithromycin	Floating gels	Antibacterial	Gastric Ulcers	Dipen et al., 2011
Cimetidine		H_2 antagonist	Peptic Ulcer	Jayswal et al., 2012
Levofloxacin		Antibacterial	Anti-*Helicobacter pylori*	Rajalakshmi et al., 2011
Metformin	Beads	Hypoglycemic	Diabetes	Vijan et al., 2012a
Propranolol		B-Blocker	Hypertension	Maiti et al., 2016
Glipizide		Hypoglycemic	Diabetes	Maiti et al., 2015
Indomethacin	Ophthalmic gel (*in-situ*)	Anti-inflammatory	Eye inflammation, Uveitis	Prajapati et al., 2013
Moxifloxacin	Ophthalmic (*in-situ* gel system)	Anti-bacterial	Ophthalmologic disorders	El-Laithy et al., 2011
Carbamazepine	Oral Gel (*in-situ*)	Anti-epileptic	Epilepsy	Prakash et al., 2014
Ascorbic acid	Gum film	Antioxidant and nutritional supplement	Food Industry	Prajapati et al., 2013
Carvedilol	Microsphere (Hydrogel)	Antihypertensive	Angina, Hypertension	Prajapati et al., 2013
Metoclopramide	Microsphere (Intranasal)	Prokinetic, Anti-emetic	Nausea & Vomiting in Cancer therapy, Pregnancy.	Vashisth et al., 2015
Mometasone	Nasal gel (*in-situ*)	Corticosteroid anti-inflammatory	Rhinitis, Allergies	Altunta et al., 2017

1. **Solid Orals**: A balance between the erosion and swelling or diffusion using different compositions and concentration promotes gellan to be used as a disintegrating agent for immediate release or matrix sustained release agent. The higher extent of swelling with the prolonged release is associated with the higher polymer content and vice versa. The swelling rate was found to be higher in the simulated intestinal fluid as compared to the simulated gastric fluid (SGF). A study revealed prolonged metformin release when acrylamide grafted gellan was tested in phosphate buffer (pH 6.8). A different effect was observed for ephedrine in hard gelatin capsules containing gellan. Where the formulation was tested in SGF (pH 1.5) and simulated intestinal fluids (SIF) (pH 7.5), the greater release was observed at higher pH (Osmalek et al., 2014).

2. **Beads and Capsules**: Natural polymer-based beads and capsules have drawn considerable attention and are obtained through ionotropic gelation. Dropwise addition of gellan solution from a needle into aqueous ionic solution with constant stirring results in the formation of 3-D crosslinked network of polymer. Most of the cases utilize Ca^{2+} as a cross-linking agent. Drug gets incorporated into the core, which can be dissolved initially in the polymeric solution or added later. The size, shape, and surface of beads can be optimized by adjusting the temperature, and the needle bore size, polymer concentration, pH, the rate of stirring, drying time and the rate of addition of counter-ions. In a few observations, porous beads were observed when pH was adjusted to acidic while formulating the beads and smooth surface on adjusting it to alkaline. Higher entrapment was also observed at higher pH. The resulting porous beads were stable in the acidic pH and disintegrated at a faster rate at higher pH, and this showed that gellan produced the pH sensitivity to beads and capsules (Prezotti et al., 2014). These observations are in contrast to the drugs having hydrophobic nature, which show higher release in acidic dissolution media. This was proposed to be because of modifications in gel structure on the change in H^+ levels, an increased level of H^+ compete with complexing agent, i.e., Ca^{2+}, and a change in the structure of gel are evident from higher drug release. This release may then be prolonged by increasing hydrophobicity of the core, by introducing oils (peanut or linseed) or clays in the formulations.

The gelation is also dependent on the nature of drug viz. cationic drugs may hinder the gelation by reducing the availability of counter-ion to complex the anionic heads of gum. The stability and drug release of gellan beads can be increased by mixing or coating with different polymers. Adding different polymers can enhance the physicochemical properties of gellan viz. mucoadhesiveness can be increased by adding anionic polymers as in chitosan (CS). Other reports by Kulkarni et al. (2013), for diltiazem HCl microcapsules based on interpenetrating polymer network (IPN) also show similar outcomes. Also, Ahuja et al. (2013), reported an increase in drug entrapment and mucoadhesion by using carboxymethylated gellan (Osmalek et al., 2014).

3. **In-Situ Gelling Systems**: The possibility of thixotropic transitions in low-acyl grade has led to an idea of oral *in-situ* forming gels which are considered as easily swallowable. Thus, administering gellan orally leads to the development of sustained release form in the stomach. This application has been exploited to express a feeling of fullness in treating obesity-related feeding potential. Cations may be added to delay gelling before administration. Delayed drug release was observed with increased polymer concentration. A study revealed the release of paracetamol (PM) loaded gellan gels, to be diffusion controlled. Adding calcium carbonate induced floating character in the amoxicillin system for the eradication of *H. pylori*.

4.7.2 APPLICATIONS IN OPHTHALMIC FORMULATIONS

Limitations in the corneal permeability lead to its lower absorption. Thus, improvement in bioavailability is made by the addition of polymers. Prevalence of traditional formulations as easy to administer and is well-tolerated leaves hindrance in the area of newer technologies. Also, before administering, one has to consider the possibility of removal or dilution of the drug by tear and also nasolacrimal drainage has to be surpassed to increase the bioavailability. Use of gellan as a thickening agent improves the sensitivity of eyes to higher concentrations. Gellan as a part of *in-situ* gelling systems seems to be prolonging the drug release. Already a marketed formulation using gellan named Timoptic XE® is widely accepted and recognized, it enhances the bioavailability three to four times and reducing the systemic side effects. Mucoadhesive properties were evaluated for gellan as well as its combination with alginate sodium and carboxymethylcellulose by Kesavan et al. (2016). In-vivo studies of this formulation showed that used polymers were suitable to extend the active drug retention. Shear thinning behavior was also observed and found favorable in ophthalmic delivery when tested in the presence of an artificial tear. A microemulsion based approach was followed by Tayel et al. (2013) for the delivery of Terbinafine. HCl, where gellan was used to stabilize the microemulsion. Also, gellan helps to form an O/W type emulsion because of its hydrophilic nature.

4.7.3 APPLICATIONS IN NASAL FORMULATIONS

Nasal drug delivery is associated not only with the local treatment including nasal congestion, allergic rhinitis, infections, etc., but systemic actions can also be produced for drugs which are metabolized quickly or are less available due to pre-systemic GIT metabolism. Being a well-tolerated and easily accessible site, targeting is also easy. Higher absorption, typically more or less comparable to intravenous, can be obtained because of a greater number of blood vessels. The major requirement to efficiently control the release is to induce bio-adhesive properties using specific polymers, thus giving adherence to mucosal layers. Scientists have studied the trans-mucosal delivery of higher molecular weight fluorescein dextran. Improvement in epithelial transport is seen more efficiently than the solution of isotonic mannitol. No such noxious overt responses were observed, and mean residence time was improved. An experiment conducted by Cao et al. (2007) evaluated the gellan gum as a carrier for scopolamine with the different proportion varying from 0.2 to 1%. The results showed higher bioavailability for gellan gels compared to the oral and subcutaneous route. Gellan was also investigated for microparticles loaded spontaneous gel and reveled slow release with non-Fickian mechanism (Mahajan et al., 2009). Shah et al., (2010) also conducted similar experiments with microspheres containing Sildenafil citrate, but this experiment reported a contrary release. A comparison-based study on thiolated and untreated gellan was also performed using dimenhydrinate to check for increased mucoadhesiveness (Mahajan et al., 2009).

4.7.4 GELLAN-BASED OTHER FORMULATIONS

Various experiments were performed by Coviello et al. (2007) on gellans chains which were linked to L-lysine ethyl ester moieties, and thus hydrogels were obtained, as a result, the polymer remained in a chaotic state and led to an improvement in its capacity to uptake water. In a study of related experiments shown by D'Arrigo et al., (2012) it was proposed that the cross-linking between the gellan chains resulted in increasing the extent of bioavailability of drugs low at water solubility. The carboxylic acid moieties of the polymer were linked with prednisolone and revealed the formation of *nanohydrogels*, which were biocompatible and possessed the ability to self-aggregate in an aqueous media.

Thermo-responsive microparticles of atenolol were designed by Mundargi et al. (2010), consisted of gellan gum and poly(N-isopropylacrylamide) and the two semi-IPNs with ionic cross-linking. The drug release was found to be controlled and temperature dependent. As with NIPAAM matrix, the thermos-dependent release was obtained in a fashion that at 25°C drug releases was found higher than at 37°C. This was proposed to be the temperature dependent swelling of the polymer. In 2006, Alupei et al., employed a method of solvent evaporation and made fused gellan and poly (*N*-vinyl imidazole) membranes, which showed stability and interpolymer complexation in acidic medium. Unsaturated esters in series were obtained when gellan was treated with acrylic acid, maleic anhydride or acryloyl chloride and this led to the formation of derivatives, which were able to get polymerized. The copolymerization of obtained esters and N-isopropyl acrylamide resulted in the formation of hydrogels, which could impart thermosensitive properties (Morris et al., 2012; Prajapati et al., 2013).

Yadav et al., (2014) proposed that the mucoadhesive properties of gellan could be increased by esterifying the polymer chains with thioglycolic acid which leads to decrease in the sensitivity towards ion gelation though retained its biocompatibility (Yadav et al., 2014). In 2015, Posadowska et al., demonstrated improvement in mechanical properties and on-site availability of vancomycin loaded poly(L-lactide-co-glycolide) (PLGA) nanoparticles by suspending these into gellan gum matrix. The system was found to be suitable as a treatment for osteomyelitis condition caused by *Staphylococcus* spp. The problem of on-site availability and release kinetics was overcome by gel system of gellan. Improvements in rheological properties suggest the development of injectables compared to the conventional demand for surgical implants (Posadowska et al., 2015).

4.8 APPLICATIONS IN BIOMEDICAL FIELD

4.8.1 TISSUE ENGINEERING

Recent advancements in the field of tissue engineering have opened new windows for cartilage tissue reconstruction, to overcome the limiting factor of slow self-repair. Thus, newer biomaterials are under investigation, which has the potential to swap tissue damage. Current trends have exploited gellan as carriers for various injectable homologized cells like that of bone marrow, adipocytic stem cells, etc. (Oliveira et al., 2010 a, b, c). Promising results were found in the storage of polymer matrix (Oliveira et al., 2010a). Administering gellan subcutaneously also confirms its biocompatibility and its potential to mix with the surroundings. This was confirmed by the existence of extracellular matrix components inside the implants (Osmalek et al., 2014). Also, there was no evident change in the mechanical attributes of the scaffold, which was accompanied with a decrease in the mass of it. The proposed mode could be resorption of the backbone polysaccharide and cell migration to surrounding tissue. Oliveira et al., (2010b, c.) reported the movement of animal cells was not hindered when an injectable system was tried for autologous cells in rabbits and filling of the wounded region were observed. Owing to poor mechanical properties viz. brittleness, hardness, and lower strength and its higher gelling temperature, efforts are desired to rectify these. These may include the addition of another polymer blend or covalent modification. An effort to reduce the polymer to smaller fragments by oxidation using sodium periodate was carried out by Gong et al. (2009). Also, the relationship between gelation and reaction time with sodium periodate was also proved. Histological examination evidenced that no such morphological changes and viability found which further confirmed with good proliferation during experimental duration for 150 days. Also, it was studied by Tang et al. (2012) that damage by oxidation is evident to points of crosslinking in gellan and contributes to faster degradation and water absorption (Osmalek et al., 2014).

4.8.2   TREATING INTERVERTEBRAL DISORDERS

Gellan is also considered suitable for correcting intervertebral disc disorders, especially correcting *nucleus pulposus*, the disorder may impair the quality of a patient's life due to incurred pain. Presence of non-covalent bonds in the backbone of the polymer has been ascribed as a reason of the dissolution of gellan in the physiological fluids. But original rheological ability is maintained most profoundly for gellan (Jahromi et al., 2011). Coutinho et al. (2010) worked on covalent modifications and mixing of low and high acylated gellan polymer to study its mechanical properties. Photo-crosslinked and methacrylated gellan indicated density improvements on modifications (Silva-Correia et al., 2011). The use of this polymer for intervertebral discs was proposed because of reported similar density to *nucleus pulposus*. Methacrylated congener was found to prevent angiogenesis by acting as a barrier in *nucleus pulposus* (Silva et al., 2012). Methacrylated modifications of gelatin and gellan were crosslinked, and interpenetrating networks were used to encapsulate human fibroblasts in hydrogels. These hydrogels were then subjected to photo-crosslinking, which then led to a reduction in cell viability due to aggressive conditions. Still, 70% cells retained cellular viability for three days of incubation. These grafts thus help in reducing the symptoms of the intervertebral disorders. Because of similar rheological properties, this polymer can be employed as a natural substitute for the conditions where the replacement of the soft tissue is desired.

4.8.3   GUIDED BONE REGENERATION (GBR)

This begins with the isolation of the fractured part from the side lying surrounding tissue to prevent the development of connective tissue is done using some non-degradable and degradable materials as films. Also, these are removed after recovery of bone. Delivery of living cells using the microspheres of gellan and gelatin to the damaged tissue was demonstrated by Wang et al. (2008). Having a high similarity profile as collagen and possess similar receptor sites as in fibronectin, it is now well established to improve the cellular affinity of microspheres (to fetal osteoblasts and dermal fibroblasts) by surface attachment. Both these cases demonstrate good cellular viability and proliferation with retention in morphology were seen. Stiff microgels of gellan were prepared using photo-crosslinking by Shin et al. (2014). Higher mechanical strength was observed as compared to simple gelatin-gellan double network gels when these microgels were immersed in a modified gelatin solution, and photo-crosslinking was also carried out for the same. The protoplast cells showed higher viability in gelatin as compared to gellan. A bio-absorbable film was developed by Chang et al., (2010), which was used to cover artificially induced defects of bone for two months. The idea behind this was the same, to prevent the covering of connective tissue to the damaged area. This material had a desired biodegradability, and inflammation was not observed in the surrounding tissue. It can also be used to improve the mechanical properties of materials used for reconstruction. Mechanical strength was also targeted by using complex scaffolds comprising of hydroxyapatite and gelatin by Barbani et al. (2012). Dense microstructures can be fabricated to induce specific interaction between carboxylate and amides groups in gellan to improve its Young's modulus. The gelatin-hydroxyapatite and gelatin-gellan interactions were confirmed and found similar to the matrix as in natural bone using spectral analysis (Douglas et al., 2014).

4.8.4   SURGERY AND WOUND HEALING

Recent searches have shown that the gellan gum is widely used for wound healing dressings due to its non-toxic properties and biocompatibility. A series of the water-insoluble film (possessing anti-adhesive properties) of low- acyl gellan, 26μm thick, crossed linked with EDC were prepared by Lee et al., (2010) which did not exert any toxic effects on blood and fibroblasts cells. During *in-vivo* studies on rats after the surgical excision, a slight inflammation was marked giving good results in the long run. Lee et al. (2012) designed an anti-adhesive type of material in which gellan chains were grafted with cinnamate moieties to ensure the cross-linking process and photosensitivity required to obtain anti-inflammatory effects. Biocompatibility studies on properties hydrogels consisting up to 2% low acylated gellan and 3% sulfated hyaluronic acid for the evaluation of gels in epidural post-surgical scar prevention showed that gellan had excellent elastic properties with no hemolytic effect and hydrogel was stabilized up to 12 months (Osmalek et al., 2014).

In 2012, a silver release wound dressing was designed by Cencetti et al. (2012), possessing anti-microbial properties. The release of silver was enhanced by using PVA/ Borax system. Various patches mainly non-woven gellan-hyaff, which are derivative (benzyl) of hyaluronic acid is used as a matrix.

4.8.5 GELLAN IN OTHER BIOMEDICAL AREAS

The gellan is used as a material for the preparation of dental cavity fillings after tooth extraction. Gellan concentrations in the range of 0.75–1.75% were evaluated for sponges preparation which was further cross-linked with EDC for 24 hours and various observations were made that this material possessed a high stability *in-vitro* with a higher rate of absorption and increased pore diameter depending upon the content of gellan (Prajapati et al., 2013).

Gellan sulfate is highly used for the treatment of rheumatoid arthritis as they bind to fibronectin molecules. Some other uses of gellan sulfate include the development of cell hybrid materials, which were used for artificial veins design. It also possesses an anticoagulation action. It was also shown that gellan could also be used as a device component for gene delivery and non-viral trans-infecting agents like branched polyethyleneimine (PEI) (Goyal et al., 2011).

4.9 FUTURE PERSPECTIVES

Due to exploitable potential, gellan has diverse opportunities to be used as bio-ink. Gellan as bio-ink could be applied in cartilage 3-D printing using extrusion-based hydrogel bioprinting technique. There it possesses an area of immense opportunity for the regeneration of chondrocytes specifically, Auricular, and nasal. Improvement in mechanical properties may be seen to reinforce the scaffold, to make it applicable to the bone tissue engineering. As previously described these scaffolds for adipocytic stem cells are good carriers, these adipocytic stem cells grow well into spongy scaffolds. The reinforcement may be carried out using advanced bioactive materials such as a type of glass in combination with gellan scaffold (Gantar et al., 2014). Other are modified gellan based nanoparticles by improvement in stability and texture. The fabrication may be desired due to the raw properties of gellan. In the coming future, increased research may be seen resulting in possible marketed products as well.

4.10 CONCLUSION

Extensive studies on a polysaccharide of microbial origin, have led to its development as a multifunctional adjuvant in various formulations. Also, being rich in advantageous properties viz. biocompatibility, cation induced gelation, hydrogel formation, biodegradability, mucoadhesiveness has made it an inevitable excipient for ophthalmic, nasal, oral, and many other formulations. The exploration is not only limited to the pharmaceutical field, but biomedical areas are also being explored as in tissue engineering, intervertebral disc-implants, wound healing, etc. The work done on the application part is still under laboratory evaluation, and so far, no marketed formulation is available.

KEYWORDS

- biopolymer
- bio-polysaccharides
- drug delivery
- gellan
- natural polymers

REFERENCES

Ahuja, M., Singh, S., & Kumar, A., (2013). Evaluation of carboxymethyl gellan gum as a mucoadhesive polymer. *International Journal of Biological Macromolecules, 53*, 114–121.

Akifuddin, S. K., Abbas, Z., Marihal, S., Ranadev, A. K., Kumar, I. H., & Kulkarni, R., (2013). Preparation, characterization and *in-vitro* evaluation of microcapsules for controlled release of diltiazem hydrochloride by ionotropic gelation technique. *Journal of Applied Pharmaceutical Science, 3*(4), 035–042.

Altunta, E., & Yener, G., (2017). Formulation and evaluation of thermoreversible *in situ* nasal gels containing mometasone furoate for allergic rhinitis. *AAPS Pharm. Sci. Tech.*, 1–10.

Alupei, I. C., Popa, M., Bejenariu, A., Vasiliu, S., & Alupei, V., (2006). Composite membranes based on gellan and poly(N-Vinylimidazole): Synthesis and characterization. *European Polymer Journal, 42*(4), 908–916.

Bacelar, A. H., Silva-Correia, J., Oliveira, J. M., & Reis, R. L., (2016). Recent progress in gellan gum hydrogels provided by functionalization strategies. *J. Mater. Chem. B*, *4*(37), 6164–6174.

Bhimani, D. R., Patel, J. K., Patel, V. P., & Detroja, C., (2011). Development and evaluation of floating in situ gelling system of clarithromycin. *IJPER*, *3*(5), 32–40.

Cencetti, C., Bellini, D., Pavesio, A., Senigaglia, D., Passariello, C., Virga, A., & Matricardi, P., (2012). Preparation and characterization of antimicrobial wound dressings based on silver, gellan, PVA and borax. *Carbohydrate Polymers*, *90*(3), 1362–1370.

Coutinho, D. F., Sant, S. V, Shin, H., Oliveira, J. T., Gomes, M. E., Neves, N. M., Khademhosseini, A., & Reis, R. L., (2010). Modified gellan gum hydrogels with tunable physical and mechanical properties. *Biomaterials*, *31*(29), 7494–7502.

Coviello, T., Matricardi, P., Marianecci, C., & Alhaique, F., (2007). Polysaccharide hydrogels for modified release formulations. *Journal of Controlled Release*, *119*(1), 5–24.

D'Arrigo, G., Di Meo, C., Gaucci, E., Chichiarelli, S., Coviello, T., Capitani, D., Alhaique, F., & Matricardi, P., (2012). Self-assembled gellan-based nanohydrogels as a tool for prednisolone delivery. *Soft Matter*, *8*(45), 11557–11564.

Douglas, T. E. L., Wlodarczyk, M., Pamula, E., Declercq, H. A., de Mulder, E. L., Bucko, M. M., Balcaen, L., Vanhaecke, F., Cornelissen, R., Dubruel, P., & Jansen, J. A., (2014). Enzymatic mineralization of gellan gum hydrogel for bone tissue-engineering applications and its enhancement by polydopamine. *Journal of Tissue Engineering and Regenerative Medicine*, *8*(11), 906–918.

El-Laithy, H. M., Nesseem, D. I., El-Adly, A. A., & Shoukry, M., (2011). Moxifloxacin-gelrite in situ ophthalmic gelling system against photodynamic therapy for treatment of bacterial corneal inflammation. *Archives of Pharmacal Research*, *34*(10), 1663–1678.

Gantar, A., Da Silva, L. P., Oliveira, J. M., Marques, A. P., Correlo, V. M., Novak, S., & Reis, R. L., (2014). Nanoparticulate bioactive-glass-reinforced gellan-gum hydrogels for bone tissue engineering. *Materials Science and Engineering: C*, *43*, 27–36.

Goyal, R., Tripathi, S. K., Tyagi, S., Ram, K. R., Ansari, K. M., Shukla, Y., Chowdhuri, D. K., Kumar, P., & Gupta, K. C., (2011). Gellan gum blended PEI nanocomposites as gene delivery agents: Evidences from in vitro and in vivo studies. *European Journal of Pharmaceutics and Biopharmaceutics*, *79*(1), 3–14.

Ismail, N. A., Mohamad, S. F., Ibrahim, M. A., Amin, M., & Anuar, K., (2014). Evaluation of gellan gum film containing virgin coconut oil for transparent dressing materials. *Advances in Biomaterials*, *2014*, Article ID 351248, 12 pages. DOI: doi.org/10.1155/2014/351248.

Jahromi, S. H., Grover, L. M., Paxton, J. Z., & Smith, A. M., (2011). Degradation of polysaccharide hydrogels seeded with bone marrow stromal cells. *Journal of the Mechanical Behavior of Biomedical Materials*, *4*(7), 1157–1166.

Jayswal, B. D., Yadav, V. T., Patel, K. N., Patel, B. A., & Patel, P., (2012). Formulation and evaluation of floating in situ gel based gastro retentive drug delivery of cimetidine. *International Journal for Pharmaceutical Research Scholars*, *1*(2), 327–337.

Kesavan, K., Kant, S., & Pandit, J. K., (2016). Therapeutic effectiveness in the treatment of experimental bacterial keratitis with ion-activated mucoadhesive hydrogel. *Ocular Immunology and Inflammation*, *24*(5), 489–492.

Lee, M. W., Tsai, H. F., Wen, S. M., & Huang, C. H., (2012). Photocrosslinkable gellan gum film as an anti-adhesion barrier. *Carbohydrate Polymers*, *90*(2), 1132–1138.

Mahajan, H., Shaikh, H., Gattani, S., & Nerkar, P., (2009). In-situ gelling system based on thiolated gellan gum as new carrier for nasal administration of dimenhydrinate. *Int. J. Pharm. Sci. Nanotechnol.*, *2*(2), 544–550.

Maiti, S., (2016). Engineered gellan polysaccharides in the design of controlled drug delivery systems. *In Emerging Research on Bioinspired Materials Engineering*, 268–295.

Maiti, S., Laha, B., & Kumari, L., (2015). Gellan micro-carriers for pH-responsive sustained oral delivery of glipizide. *Farmacia*, *63*(6), 913–921.

Martin, D., Steinhoff, B., Warren, H., & Clark, R., (2014). Enhanced gelation properties of purified gellan gum. *Carbohydrate Research*, *388*, 125–129.

Morris, E. R., Nishinari, K., & Rinaudo, M., (2012). Food hydrocolloids gelation of gellan E: A review. *Food Hydrocolloids*, *28*(2), 373–411.

Mundargi, R. C., Shelke, N. B., Babu, V. R., Patel, P., Rangaswamy, V., & Aminabhavi, T. M., (2010). Novel thermo-responsive semi-interpenetrating network microspheres of gellan gum-poly(N-isopropylacrylamide) for controlled release of atenolol. *Journal of Applied Polymer Science*, *116*(3), 1832–1841.

Norton, A. B., Cox, P. W., & Spyropoulos, F., (2011). Food hydrocolloids acid gelation of low acyl gellan gum relevant to self-structuring in the human stomach. *Food Hydrocolloids*, *25*(5), 1105–1111.

Oliveira, E., Assunção-Silva, R. C., Ziv-Polat, O., Gomes, E. D., Teixeira, F. G., Silva, N. A., Shahar, A., & Salgado, A. J., (2017). Influence of different ECM-like hydrogels on neurite outgrowth induced by adipose tissue-derived stem cells. *Stem Cells International*, *2017*. DOI: doi.org/10.1155/2017/6319129.

Oliveira, J. T., Martins, L., Picciochi, R., Malafaya, P. B., Sousa, R. A., Neves, N. M., Mano, J. F., & Reis, R. L., (2010). Gellan gum: A new biomaterial for cartilage tissue engineering applications. *Journal of Biomedical Materials Research Part A*, *93A*(3), 852–863.

Osmalek, T., Froelich, A., & Tasarek, S., (2014). Application of gellan gum in pharmacy and medicine. *International Journal of Pharmaceutics*, *466*, 328–340.

Posadowska, U., Brzychczy-Wloch, M., & Pamula, E., (2015). Injectable gellan gum-based nanoparticles-loaded system for the local delivery of vancomycin in osteomyelitis treatment. *Journal of Materials Science: Materials in Medicine*, *27*(1), 9.

Prajapati, V. D., Jani, G. K., Zala, B. S., & Khutliwala, T. A., (2013). An insight into the emerging exopolysaccharide gellan gum as a novel polymer. *Carbohydrate Polymers*, *93*(2), 670–678.

Prakash, K., Satyanarayana, V., Nagiat, H., Fathi, A., Shanta, A., & Prameela, A., (2014). Formulation development and evaluation of novel oral jellies of carbamazepine using pectin, guar gum, and gellan gum. *Asian Journal of Pharmaceutics*, 241.

Prezotti, F. G., Stringhetti, B., Cury, F., & Evangelista, R. C., (2014). Mucoadhesive beads of gellan gum/pectin intended to controlled delivery of drugs. *Carbohydrate Polymers*, *113*, 286–295.

Rajalakshmi, R., Sireesha, A., Subhash, K. V., Pavani, P. V., & Naidu, K., (2011). Development and evaluation of a novel floating *in situ* gelling system of levofloxacin hemihydrate. *International Journal of Innovative Pharmacy and Research*, *2*(1), 102–108.

Shin, H., & Olsen, B. D., (2014). Khademhosseini, A. gellan gum microgel-reinforced cell-laden gelatin hydrogels. *J. Mater. Chem. B.*, *2*(17), 2508–2516.

Silva, N. A., Cooke, M. J., Tam, R. Y., Sousa, N., Salgado, A. J., Reis, R. L., & Shoichet, M. S., (2012). The effects of peptide modified gellan gum and olfactory ensheathing glia cells on neural stem/progenitor cell fate. *Biomaterials, 33*(27), 6345–6354.

Tang, Y., Sun, J., Fan, H., & Zhang, X., (2012). An improved complex gel of modified gellan gum and carboxymethyl chitosan for chondrocytes encapsulation. *Carbohydrate Polymers, 88*(1), 46–53.

Tayel, S. A., El-Nabarawi, M. A., Tadros, M. I., & Abd-Elsalam, W. H., (2013). Promising ion-sensitive in situ ocular nanoemulsion gels of terbinafine hydrochloride: Design, *in vitro* characterization and *in vivo* estimation of the ocular irritation and drug pharmacokinetics in the aqueous humor of rabbits. *International Journal of Pharmaceutics, 443*(1), 293–305.

Vashisth, P., Singh, H., Pruthi, P. A., & Pruthi, V., (2015). Gellan as novel pharmaceutical excipient. In: *Handbook of Polymers for Pharmaceutical Technologies: Structure and Chemistry*, 1–21.

Vieira, S., Vial, S., Maia, F. R., Carvalho, M., Reis, R. L., Granja, P. L., & Oliveira, J. M., (2015). Gellan gum-coated gold nanorods: An intracellular nanosystem for bone tissue engineering. *RSC Advances, 5*(95), 77996–78005.

Vijan, V., Kaity, S., Biswas, S., Isaac, J., & Ghosh, A., (2012). Microwave-assisted synthesis and characterization of acrylamide grafted gellan, application in drug delivery. *Carbohydrate Polymers, 90*(1), 496–506.

Wang, C., Gong, Y., Lin, Y., Shen, J., & Wang, D. A., (2008). A novel gellan gel-based microcarrier for anchorage-dependent cell delivery. *Acta Biomaterialia, 4*(5), 1226–1234.

Yadav, S., Ahuja, M., Kumar, A., & Kaur, H., (2014). Gellan–thioglycolic acid conjugate: Synthesis, characterization and evaluation as mucoadhesive polymer. *Carbohydrate Polymers, 99*, 601–607.

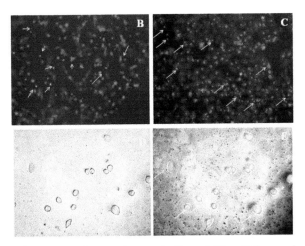

FIGURE 1.4　Fluorescent microscopic examination of C6 cell lines and neuro 2a cell lines treated with (B) RhB labeled AZT-GAAD NPs (C) merged image of AZT-GAAD NPS with cell nuclei stained blue with Hoechst 33342(D) bright field image of neuro 2a cells(E) RhB labeled AZT-GAAD NPs.

Source: Joshy et al., Copyright © 2018 with permission from Elsevier B.V.

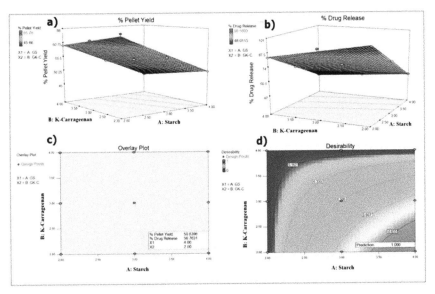

FIGURE 1.7　Response surface plots showed the influence of starch concentration (X1) and κ- carrageenan concentration (X2) on (a) % Pellet size (Y1), (b) % Drug release (Y2), (c) overlay plot, and (d) desirability plot.

Source: Sonawane et al., Copyright © 2018 with permission from Elsevier B.V.

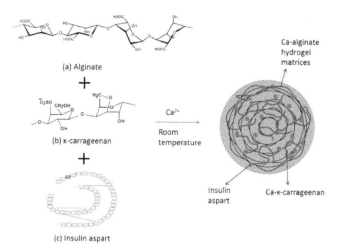

FIGURE 1.9 Schematic diagram of the formation of insulin as part-loaded alginate/κ-carrageenan composite hydrogel beads.

Source: Lim et al., Copyright © 2017 with permission from Elsevier B.V.

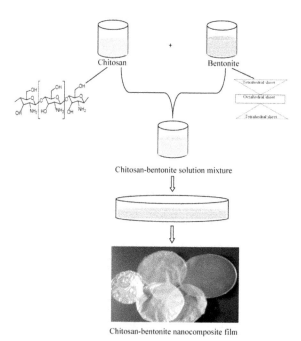

SCHEME 1.1 A schematic representation for the preparation of chitosan-bentonite nanocomposite film.

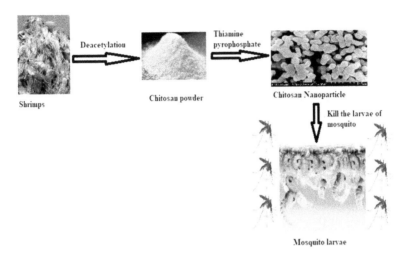

SCHEME 1.2   Schematic illustration of chitosan extraction from shrimp shells and synthesis of CS NPs from extracted chitosan using TPP as a reducing agent.

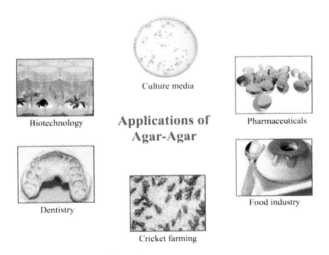

FIGURE 3.1   Potential applications of agar-agar.

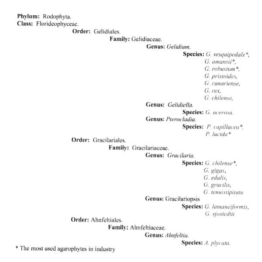

Phylum: Rodophyta.
Class: Florideophyceae.
    Order: Gelidiales.
        Family: Gelidiaceae.
           Genus: *Gelidium*.
               Species: *G. sesquipedale*,*
                      G. amansii*,
                      G. robustum*,
                      G. pristoides,
                      G. canariense,
                      G. rex,
                      G. chilense,
            Genus: *Gelidiella*.
               Species: *G. acerosa*.
            Genus: *Pterocladia*.
               Species: *P. capillaceu*,*
                      P. lucida*
    Order: Gracilariales.
        Family: Gracilariaceae.
           Genus: *Gracilaria*.
               Species: *G. chilense*,*
                      G. gigas,
                      G. edulis,
                      G. gracilis,
                      G. tenuistipitata
           Genus: Gracilariopsis
               Species: *G. lamaneiformis,*
                      G. sjostedtii
    Order: Ahnfeltiales.
        Family: Ahnfeltiaceae.
           Genus: *Ahnfeltia*.
              Species: *A. plycata*.

* The most used agarophytes in industry

FIGURE 3.3 Different species of the agarophytes.
Reprinted with permission. Copyright reserved by Elsevier (Armisén and Galatas, 2009).

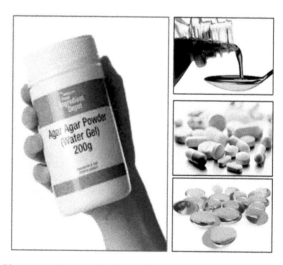

FIGURE 3.5 Pharmaceutical applications of agar-agar.

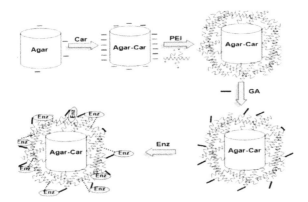

FIGURE 3.6 Illustration showing steps in preparing and activating agar-car gel disk for immobilization of enzymes.
Reprinted with permission, Copyright reserved by Springer (Wahba and Hassan, 2017).

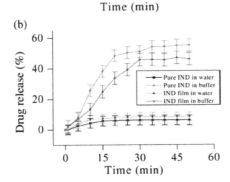

FIGURE 5.1 Release profiles from (a) pure paracetamol (PM) and PM-loaded films, (b) pure indomethacin (IND) and IND-loaded films in two different pH media (deionized water and buffer solution pH 6.2).
Reprinted with permission from (Kianfar et al., 2012), Copyright © 2012 Taylor & Francis Group.

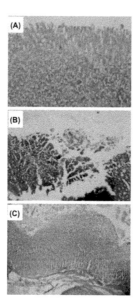

FIGURE 5.2   Stomach histopathology of control rats (A), rats treated with pure ketoprofen (B), and rats treated with interpenetrating network hydrogel beads (C). (Hematoxylin-Eosin, 100×.). It is reprinted from (Kulkarni et al., 2012), Copyright 2012, with permission from Elsevier.

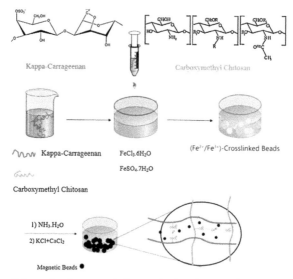

FIGURE 5.3   Schematic diagram of the synthesis of magnetic κ-carrageenan/chitosan beads. Reprinted from (Mahdavinia et al., 2015), Copyright 2015, with permission from Elsevier.

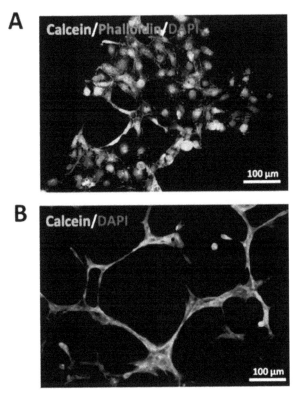

FIGURE 5.4 Cells that were retrieved after 21 days of culture in fibers. DAPI staining is used for DNA and phalloidin for F-actin. Green-calcein represents live cells.
Adapted with permission from (Mihaila et al., 2014), Copyright 2014 American Chemical Society.

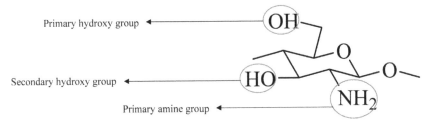

FIGURE 6.3 Reactive sites for chitosan modifications.
Source: Agarwal et al., 2015, with permission from Journal of Pharma Research

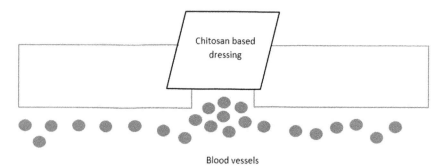

FIGURE 6.5   Schematic representation of the working of the chitosan-based wound dressing.

Source: Cheung et al., 2015, Free Access.

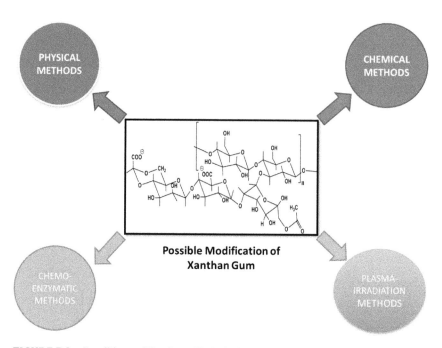

FIGURE 7.2   Possible modification of XG chains used in pharmaceutical applications.

CHAPTER 5

Pharmaceutical Applications of Carrageenan

A. PAPAGIANNOPOULOS and S. PISPAS

Theoretical and Physical Chemistry Institute, National Hellenic Research Foundation, 48 Vassileos Constantinou Avenue, 11635 Athens, Greece

5.1 INTRODUCTION

The natural polysaccharide carrageenan is extracted from edible red seaweed, and it consists of galactose and anhydrogalactose units, which are connected by alternating α-(1,3) and β-(1,4) glycosidic bonds (Malafaya et al., 2007). Its main types κ (kappa), ι (iota) and λ (lambda) are distinguished by the sulfate groups number per disaccharide repeating unit (one, two or three) and hence they have different linear charge density (κ < ι < λ) and solubility (Raveendran et al., 2013; Sosnik et al., 2014). They can form rigid networks, soft gels, and viscous fluids and offer a great variety of possible applications, as they will interact electrostatically with proteins and charged drugs.

Carrageenan is customarily used to provoke inflammation in order to test the anti-inflammatory activity of several substances (Hadj et al., 2015). Additionally, it has been employed in pharmaceutical formulations for prolonged release of drugs (Picker, 1999a; Picker, 1999b). Recent investigations have focused on the production of carrageenan materials for pharmaceutical applications in many cases. The potential for drug delivery at the nanoscale has been demonstrated by protein-loaded nanoparticles (Dionísio et al., 2013) and liposomal formulations (Biruss and Valenta, 2006), while carrageenan has also been included in tablet formulations for sustained drug release (Hariharan et al., 1997). Notably, carrageenan finds application in mucoadhesive systems for vagina drug delivery (Caramella et al., 2015) and in particular its anti-HIV activity has been proposed as an additional feature on top of its use as an excipient (Vlieghe et al., 2002). In this chapter, the recent developments in the potential of carrageenan for pharmaceutical applications are presented. Examples include its bioadhesive properties in nasal drug delivery (Song and Eddington, 2012), vaccine delivery (Bertram et al., 2010) and administration with buccal films (Kianfar et al., 2012). Research on hydrogels for stimuli-responsive substance release (Alvarez-Lorenzo et al., 2013), wound healing (Silva et al., 2011) bionanocomposites (Khan et al., 2017) and tissue engineering (Mihaila et al., 2013) is included.

5.2 ANTI-HIV ACTIVITY

Carrageenan possesses anti-HIV activity as it inhibits entry and attachment of the virus (Woodsong and Holt, 2015) and has been the basis of carraguard. It is one of a series of polymers considered as mucoadhesive materials while it particularly prevents infection by binding to the virus surface, stopping microbes from adhering to cells (Valenta, 2005). The polysaccharide has also been involved in carrier gels (Singer et al., 2011) for non-nucleoside reverse transcriptase inhibitor (NNRTI) MIV-150. Carrageenan gels were shown to be safe (Skoler-Karpoff et al.), and their physicochemical properties render them promising for the delivery of anti-HIV active pharmaceutical ingredients (APIs) (Fernández-Romero et al., 2007, 2012). Additionally, carrageenan has been applied in microbicides with antiviral activity (Maguire et al., 1998) and animal models against HPV (Roberts et al., 2011). Acetate-carrageenan is under investigation for possible activity against HSV and HPV (Levendosky et al., 2015).

Carrageenan is met in applications with potential in HIV (human immunodeficiency virus) prevention and other infections that are sexually transmitted. Vagina is characterized as an excellent way to deliver nucleic acids, proteins, and other therapeutic macromolecules, and hence, intravaginal drug administration is generally acceptable. Biodegradable films to deliver siRNA (small interfering RNA) to the vaginal mucosa have been proposed since siRNA is considered as an effective compound to reduce viral entry into cells (Gu et al., 2015). Biodegradable films made of poly(vinyl alcohol) and λ-carrageenan were homogeneously loaded by siRNA containing nanoparticles. The administration of the film would be followed by its disintegration in the vaginal lumen and the subsequent delivery of nanoparticles of siRNA to the HLA-DR+ dendritic cells.

Polymeric carriers originating from κ-carrageenan conjugated with 3'-azido-3'-deoxythymidine (AZT) were considered for the production of an anti-HIV drug capable of exploiting the anti-HIV properties of both of its components in a synergistic manner (Vlieghe et al., 2002). κ-Carrageenan is active against enveloped viruses like other sulfated polysaccharides (Schaeffer and Krylov, 2000) and has itself anti-HIV activity (Baba et al., 1988). Conjugates were tested on MT-4 cells before or after infection with HIV-1 virus. The synergistic activity of κ-carrageenan and AZT was obtained at a critical ratio of AZT as it was found on drug-infected MT-4 cells treated by the conjugate drug (Vlieghe et al., 2002).

A flux controlled pump for controlled release of macromolecules in the female genital tract has been proposed and tested *in vivo* on female rabbits (Teller et al., 2014). The pump administered the macromolecules in the form of pellets through orifices. The release rate was modified by the swelling polymer, e.g., carrageenan, and orifice size.

In vaginal drug delivery, physiological mechanisms that are responsible for drug removal and restricted residence time in the lumen have to be overcome. In order to extend the duration of drug persistence in the vaginal cavity, mucoadhesion and thermogelling are two distinct approaches (Caramella et al., 2015). Carrageenan has been mixed with Poloxamer® in order to produce thermoresponsive drug carriers with the sustained release (Liu et al., 2009).

5.3 ANTI-INFLAMMATORY ASSAYS

Carrageenan is often used as an agent of controlled inflammation in tissues for the evaluation of natural and synthetic pharmaceutical compounds. The injection into the air-pouch of rat back demonstrated the anti-inflammatory effect of *Cynodon daktylon*. The plant extract acted against acute inflammation and increased angiogenesis. Hence *C. dactylon* was proposed as a potential angiogenic compound based on the enhanced expression of vascular endothelial growth factor (Soraya et al., 2015). The activity of Eunicella singularis extract against inflammation in reference to acetylsalicylate of lysine was tested by paw edema induced by carrageenan. The rat paw edema development caused by carrageenan is related to the acute phase inflammation events interceded by histamine, bradykinin, and prostaglandins that are produced under the influence of cyclooxygenase-2. The authors concluded (Deghrigue et al., 2014) that *E. singularis* activity was possibly due to blocking the cyclooxygenase-2.

Synthetic substances as imidazolyl-1,3,4-oxadiazoles and 1,2,4-triazoles were proposed for anti-inflammation applications in an attempt to produce drugs with reduced side effects in long-term clinical employment. Their anti-inflammatory profile was tested by carrageenan-induced rat paw edema. In brief, the edema was quantified by the difference in thickness between saline and carrageenan affected paw. Similarly, activity against inflammation was evaluated as the inhibition percentage in comparison to the control group. Although the activity was similar to the one of standard indomethacin, the produced compounds revealed a safer profile (Almasirad et al., 2014). Sulfated seaweed polysaccharides were screened with carrageenan as a phlogistic agent after their physicochemical characterization. Sulfated polysaccharides from *C. sedoides*, *C. compressa*, and *C. crinita* inhibited the formation of the edema significantly, 3 h after the injection of carrageenan (Hadj et al., 2015).

Potential dual drugs, i.e., dual inhibitors for the treatment of inflammation conditions, showed advantages of gastrointestinal safety and antioxidative effects. In vitro and in vivo studies put forward LQFM-91 as a dual inhibitor of cyclooxygenase-5/ lipoxygenase and a compound with the ability to reduce edema, leukocyte migration and inflammatory cytokines (Zhang et al., 2012). Mechanical nociceptive thresholds were determined after intraplantar injection into the right hind paw of *Swiss* albino mice under the paw pressure test (Kawabata et al., 1992). Saline was used as a control injection into the left hind paw. Acute inflammation was modeled by edema on the hind paw (Winter et al., 1962) where the inflammatory response was determined by measuring the volumes of the two paws, i.e., carrageenan-injected and saline injected (control) paw. Induced pleurisy was realized by the introduction of carrageenan into the pleural cavity the treatment with the potential drug and counting of the number of leukocytes (Vinegar et al., 1973). 2,4-Dichlorophenoxy acetic acid, a substance customarily employed in tissue culture of plants, has recently been proposed as a new anti-inflammatory agent. Based on *in vivo* edema tests with carrageenan (Khedr et al.,

2014) the action of the drug in a gel formulation was found similar to the one of ibuprofen.

Orally administered LYSO-7 a new potential anti-inflammatory drug with hybrid features has shown high *in vivo* anti-inflammatory potential in comparison with other drugs. Inflammation was experimentally induced by carrageenan solutions injected subcutaneously on mice dorsal surfaces creating air pouches (Santin et al., 2013). LYSO-7 was able to reduce migration of neutrophil to the affected area with doses that were lower than the ones for indomethacin, celecoxib, or pioglitazone. Thermal hyperalgesia model supported the potential of newly synthesized compounds with the reference analgesic drug Ketorolac measuring the paw withdraw latency (Abdelazeem et al., 2014). Discrete oligoethyleneglycolation (dPEGylation) strategy was applied on neuropeptides as galanin and neuropeptide Y in order to create periphery selective analogs and prevent penetration into the central nervous system. The dPEG analogs lacked apparent antiseizure activities keeping the analgesic properties. Carrageenan inflammatory pain models after systematic administration were exploited to test analgesic activity (Zhang et al., 2012).

Treatment of ulcerative colitis with salicylazosulfapyridine and prednisolone was investigated after adverse symptomatology caused by carrageenan in an early study (Boxenbaum and Dairman, 1977). In this work, the researchers provided the polysaccharide to guinea pigs with drinking water and concluded that the tested model might not be effective for screening therapeutic agents for human ulcerative colitis conditions. Another model for carrageenan-induced inflammation to rat intestine involving oral administration was proved useful for the investigation of human inflammatory bowel disease (Moyana and Lalonde, 1990).

5.4   LIPOSOMAL FORMULATIONS

Hydrophilic anticancer chemotherapeutic drugs can be delivered by loading on liposomes. Lipid-carrageenan hybrid nanoparticles have been designed for the delivery of drugs against cancer, i.e., mitoxantrone hydrochloride (HCl) (Ling et al., 2016). The cationic drug was electrostatically coupled with carrageenan providing increased encapsulation. The combination of hydrophobic/electrostatic interactions and the structure of hybrid liposomes resulted in sustainable drug release. Apparent bioavailability in rats was more than three times higher than in solutions of pure drug. Additionally, the relative cellular association in breast cancer resistance cells was more than nine times higher. Similarly, the cytotoxicity of mitoxantrone was significantly enhanced after encapsulation (Ling et al., 2016).

It has been reported that drug permeability through the stratum corneum of the skin is improved by the use of liposomes. Topical formulations of steroid hormones are meant for several therapeutic needs and are beneficial for metabolism (Biruss and Valenta, 2006). The lipophilic properties and function of steroid hormones vary although their binding mechanisms may be similar. It has been postulated that the interactions of liposomes with the intracellular lipids in the skin are responsible for their increased penetration (Šentjurc et al., 1999). Forming layers of polymers around the vesicles is a way to enhance stability against microbial and chemical loads for drug-loaded liposomes. For example, carrageenan and polyacrylic acid derivatives introduce anti-microbial activity. Additionally, the rheological properties of the formulations can be adjusted by the presence of polymers. Liposomal formulations made of DPPC (1,2-dipalmitoyl-sn-glycero-3-phosphocholine) with incorporated hormones as 17-β-estradiol, progesterone, CPA, and finasteride were investigated in terms of chemical stability and skin permeation (Biruss and Valenta, 2006). Coating of liposomes with carrageenan increased permeation rate for most of the hormones.

5.5 TABLET FORMULATIONS

Carrageenan has found application in the oral release of extended-release tablets and extrusion aid in the pellet production (Li et al., 2014). Its negative charge and gelling behavior provide the ability for viscosity modification and controlled release of therapeutic substances. In wet extrusion/spheronization technology (Yoo and Kleinebudde, 2009), carrageenan may act as a pelletization aid. It has also been proposed as an alternative to the standard extrusion aid microcrystalline cellulose, which has some drawbacks such as pellet non-disintegration (Basit et al., 1999). Release of drugs from microcrystalline cellulose-based pellets was controlled by diffusion. In κ-carrageenan pellets, the rapid disintegration led to rapid release (Kranz et al., 2009).

Microcrystalline cellulose binding properties and ability to retain water are the main reasons why it has been considered a superior excipient in extrusion-spheronization for a long time (Valle et al., 2016). As a number of drugs are unstable or incompatible with microcrystalline cellulose, other polymeric substances have been proposed for its replacement. κ-Carrageenan is promising for such applications because of the fast disintegration of its pellets, which is critical for substances with poor solubility in water. Hence it has been used as a spheronization aid along with chitosan (CS) as a diluent and Carbopol 974P as a binder. During the extrusion process plasticization caused by the ability of carrageenan to absorb water allows for spherical or almost spherical pellets (Valle et al., 2016).

Drug release from tablets has been routinely controlled by polymers. The exploitation of carrageenans as functional tableting excipients and their use in tablet formulation in practice were reported a couple of decades ago (Picker, 1999a). Tablets with satisfactory compatibility and consolidation were prepared from ι, κ, and λ-carrageenans. The viscoelastic compacts were described as soft, i.e., not to causing high mechanical stresses to the incorporated drugs. This soft inclusion was supported by the high elastic recovery of the compacts. The release behavior of the carrageenan tablets was co-evaluated with the swelling, viscoelasticity, and water sorption of the hydrocolloid matrix (Picker, 1999b). Increase in viscosity, decrease in water sorption, or decrease in swelling increased drug release. Swelling/drug release properties could be tuned by the addition of salts before tableting but only up to contents that do not cause matrix disruption. In addition, drug release depended on whether the drug was ionic or nonionic.

Angina treatment demands controlled release and absorption of drugs and reduced hepatic first-pass metabolism. Proliposomal drug delivery can enhance oral bioavailability, but release mediated by liposomes has to be sustained. κ-Carrageenan, instead of microcrystalline cellulose, co-blended with proliposomal powders resulted to pellets that would readily dissolve in water because of κ-carrageenan's hydrophilic nature and consequent water absorption properties. The methodology that was followed showed the advantages of sustained-release pellets and provided effective approaches in the oral administration of chronotherapeutic drugs for angina therapy with the delivery of protocatechualdehyde-phospholipid complex (Zhang et al., 2016b).

Carrageenan matrices have been used in formulations for the model drug anhydrous theophylline. They were prepared by direct compression and optimized in terms of physical properties and drug release profiles. The drug release properties were estimated at a content of carrageenan 40%, and two excipients lactose fast flow and emcompress (Rosario and Ghaly, 2002). The gel layer of carrageenan (two different conditions, i.e., phosphate buffer pH 7.4 and 0.1 N HCl) was affected leading to a higher release of drug in comparison to distilled water. The thickness of the hydrogel layer is also affected by the speed of rotation speed in the apparatus of dissolution. The drug release was diffusive up to 90 min and followed zero-order model onwards.

Tablets from several polysaccharides, i.e., xanthan gum (XG), β-cyclodextrin and carrageenan or their mixtures with methacrylic acid-grafted-guar were evaluated for delivery agents for the therapy of amoe-biasis. Delivery of drugs targeted to the colon needs to be stable in the small intestine and stomach conditions. Tablets with carrageenan, pectin, or β-cyclodextrin were disassembled within 24 h and had released the most of the drug in the simulated gastric or intestinal fluid (Mundargi et al., 2007). In another study, four components were included in tablet formulations for the optimization of sustained-release. Cross-linked carboxy-methylcellulose sodium and γ-carrageenan (hydrophilic polysaccharides) were combined with two different fillers (dibasic calcium phosphate and α-lactose monohydrate). Release kinetics exponent, crushing strength, and time for the release of a predefined drug percentage were the main criteria for the evaluation of the properties (Hariharan et al., 1997).

The effect of the presence in the pellet matrix of two polysaccharides instead of one on drug release was tested by mixed matrices of carrageenan and CS in several aqueous media (Bani-Jaber and Al-Ghazawi, 2005). Drug release in inter-polyelectrolyte complexes was strongly compromised by the degree of ionization of CS. In water and simulated gastric fluid (SGF) and 75/15 carrageenan/CS content, only 20% of diltiazem. HCl was released in 10 h, which was an exceptional sustained release.

Cross-linked microgels based on carrageenan about 80 μm in diameter provided smooth and stable microbeads. In brief, macrobeads and microparticles of carrageenan were produced by conventional dipping methods and emulsion techniques, respectively. Cross-linking was achieved by the agent epichlorohydrin, which is traditionally applied on polysaccharides. Absence or presence of high salt resulted in a swelling-shrinking transition with an accompanying change in volume of the order of 400%. This strong salt dependence was put forward as a very interesting feature for drug release within the gastrointestinal tract (GIT) (Keppeler et al., 2009).

5.6 NASAL DRUG DELIVERY

The delivery of therapeutic agents to mucus layers with reduced bioavailability can be increased by the combination of bioadhesive macromolecules with the tight junction permeation enhancer (AT1002). Nasal administration of inulin, calcitonin, and saquinavir was investigated under the influence of AT1002 by in Sprague–Dawley rats. The two polysaccharides, ι-carrageenan, and Na-carboxymethyl cellulose enhanced permeation in combination with AT1002 and consequently increased the absorption of inulin (molecular weight marker) and therapeutic agents with low bioavailability (calcitonin and saquinavir). This work supported the use of carrageenan for the advancement of pharmaceutical agents with reduced bioavailability to mucus (Song and Eddington, 2012).

Mucosal surfaces have been under research as alternative administration routes. Nasal mucosa ensures rapid drug absorption and large absorptive surface without the drawback of the low permeability of gastrointestinal mucosa and first pass metabolism during oral administration (Djupesland, 2013). As a bioadhesive polymer carrageenan has been employed in such applications. In situ gelling nasal gels from carrageenan/Na-alginate blends provided adjustable drug release rate (Bertram and Bodmeier, 2012). Solution viscosity and water uptake could also be tuned by the ratio of the two polysaccharides. The use of blends was more effective than adjusting the molar mass of a single polymer.

The sulfated polysaccharide ι-carrageenan has indeed demonstrated antiviral activity against respiratory viruses. It's possible to direct binding to viruses prevents the attachment of viruses to host cells (Grassauer et al., 2008; Leibbrandt et al., 2010). Clinical trials of ι-carrageenan containing nasal sprays supported the antiviral effectiveness of this polysaccharide. Two randomized, double-blind placebo-controlled trials from patients with acute common cold were analyzed. Carrageenan nasal spray was found effective as it increased viral clearance and reduced relapses of symptoms in children and adults (Koenighofer et al., 2014). It has been indicated that cold symptoms are strongly reduced during the first days when symptoms are more severe. The nasal spray appears safe, minimally invasive, and well-tolerated in trial results (Eccles et al., 2015). In conclusion, ι-carrageenan was suggested as a satisfactory option for infections of the upper respiratory tract (Ludwig et al., 2013).

5.7 VACCINE DELIVERY

The anticoagulant activity of λ- and κ-carrageenan has been demonstrated in early studies. Undegraded carrageenans were acutely toxic and combined with fibrinogen in the plasma assembled into insoluble complexes. Degraded samples from Eucheuma spinosum were less toxic and showed anticoagulant activity during intravenous injection to rabbits because they form soluble complexes with fibrinogen (Anderson et al., 1965). Influenza split vaccine contained in freeze-dried solutions of hydrophilic macromolecules were used for nasal inserts. *In situ* gelling systems had different release rates because of differences in interactions of proteins with polymers and viscosity. ι-Carrageenan, Carbopol®, and XG had similar profiles of protein release and as negatively charged polymers were dissolved without complications in the vaccine solutions (Bertram et al., 2010).

5.8 BUCCAL FILMS

Buccal films offer means to drug delivery using bioadhesion and subsequent hydration to deliver a drug across the buccal membrane. Buccal films based on carrageenan were produced and optimized for stability, drug loading, and mechanical characteristics. The transformation of the two crystalline drugs paracetamol (PM) and indomethacin to the amorphous state inside the gel film matrices offered drug stability over 12 months. Promising release profiles (Figure 5.1) were attained under saliva simulating conditions (Kianfar et al., 2012).

Lyophilization a traditional technology in the food industry is also applied to pharmaceutical products. Gels from carrageenan, poloxamer, and poly(ethylene glycol) were prepared in the form of wafers with thermal processing. The freeze-drying process included freezing, primary, and secondary drying cycles. Mechanical characterization of the wafers was performed with texture analysis (Kianfar et al., 2014). The bioadhesive properties were analyzed by attaching wafers of suitable size on the probe of a texture analyzer and measuring the force of detachment from an agar gel after hydration and adhesion of the two gels. The methodology followed may be useful for the design and optimization of buccal film drug delivery, especially because tests include swelling and hydration in simulated saliva conditions (Kianfar et al., 2014).

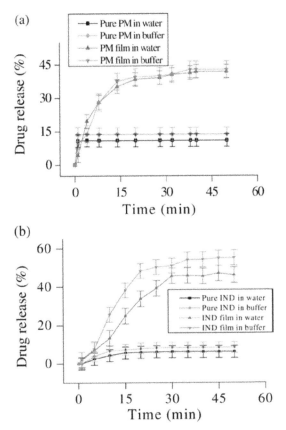

FIGURE 5.1 **(See color insert.)** Release profiles from (a) pure paracetamol (PM) and PM-loaded films, (b) pure indomethacin (IND) and IND-loaded films in two different pH media (deionized water and buffer solution pH 6.2).
Reprinted with permission from (Kianfar et al., 2012), Copyright © 2012 Taylor & Francis Group.

Stomatitis may induce ache in the oral cavity, loss of appetite, and impaired swallowing. It is a side effect of cytotoxic drugs administered for cancer chemotherapy. Mouthwashes that contain povidone-iodine are capable of preventing removal of free radicals and import of anti-neoplastic agent into the oral cavity. An allopurinol mouthwash based on ι-carrageenan and Alkox® mouthwash were investigated in terms of its potential pharmaceutical use (Hanawa et al., 2004). The authors based on adhesiveness and sensory experiments reported a mouthwash of 0.4% ι-carrageenan and 1.0% Alkox as the optimum combination.

5.9 WOUND HEALING

Cellular responses in the cornea are related to the healing process of the several possible injuries during ophthalmological operations (Netto et al., 2005). Carrageenan and gellan gum were demonstrated to enhance the delivery efficacy of antisense oligodeoxynucleotide in cornea scrape wounding (Rupenthal et al., 2011). Delivery was mostly found in the epithelium at the wound healing edges and the underlying layer of the healing. Gellan and carrageenan appeared to keep wound size better, reduce the inflammatory response, and decrease connexin protein levels. The ability of these polysaccharides for hydrogel formation in the tear fluid induced by cations was exploited. This *in situ* gel formation was considered a defining factor as it increased the precorneal retention.

Encapsulation of the hydrophobic drug ciprofloxacin in cyclic β-$(1 \rightarrow 3)$ $(1 \rightarrow 6)$ glucan/carrageenan hydrogels resulted in increased drug loading, sustained release, and antibacterial activity. Their wound healing properties were tested *in vivo* in rats. The porosity of the hydrogels induced by glucan was responsible for better cell attachment and proliferation. Healing of excision was faster in comparison to the control (Nair et al., 2016).

Gamma irradiation-induced cross-linking in polyvinyl pyrrolidone (PVP), κ-carrageenan and polyethylene glycol in aqueous solutions led to hydrogels with potential in burn wound dressing. Their physicochemical properties were comparable with the ones of commercial products while they had relatively long shelf-life (De Silva et al., 2011).

5.10 STIMULI-RESPONSIVE HYDROGELS

Ionic polysaccharides, e.g., alginate, pectin, carrageenan, gelatin, gellan, and agar are able to form hydrogels in the presence of salt ions and are commonly used as viscosity modifiers and gelling agents (Saha and Bhattacharya, 2010). Except for the aforementioned ionotropic gelation, 3D networks can form by other physical ways as creating junction zones by inter-chain double helices that can be assisted by temperature treatment. The effect of potassium and calcium anions on the structure and release properties of carrageenans was investigated by cryogenic scanning electron microscopy and Franz-cell diffusion apparatus, respectively (Thrimawithana et al., 2011). Differences in mechanical properties were attributed to the hexagonal structures found in κ-carrageenan in comparison to the rectangular pores observed in ι-carrageenan at low salt concentration. Sustained release of fluorescein was achieved from all hydrogels while a certain content of calcium caused maximal diffusivity in ι-carrageenan. Apparently, the improved tortuosity in the presence of the divalent salt was responsible for this finding (Thrimawithana et al., 2011).

Cross-linking reactions in polysaccharide hydrogels may be used in a single polysaccharide or interpenetrating networks. Glutaric acid and glutaraldehyde (GA) cross-linking in CS/κ-carrageenan hydrogel beads led to pH-responsive systems (Piyakulawat et al., 2007). The release of sodium diclofenac was sustained at pH 7.4 (small intestine condition) while it stopped at pH 1.2 (gastric condition) and 6.6 (intermediate gastric/ small intestine condition). Responsive release of diclofenac in intestinal conditions was also attained by hydrogels of κ-carrageenan with grafted polyacrylic acid. Alkaline conditions cause dissociation of carboxylic groups and lead to strong electrostatically driven swelling of the charged networks. Non-Fickian kinetics found in the release from acidic conditions was connected to diffusion coupled with polymer network dynamics. At high pH release is governed by diffusion and relaxation mechanisms (Hosseinzadeh, 2010).

Another hydrogel system that achieved controlled release in the intestine and restricted release in the acidic environment was the interpenetrating network consisting of sodium alginate and carrageenan that was grafted with polyacrylamide (Kulkarni et al., 2012). The beads of the interpenetrating hydrogel were responsive to pH and were used for delivering ketoprofen to the intestine. Several side effects of pure ketoprofen, i.e., hemorrhage; perforation, edema, and necrosis, were not found upon its encapsulation into the network (Figure 5.2). κ-Carrageenan/carboxyl cellulose beads (16–18 μm), cross-linked by genipin were produced at several relative polysaccharide ratios by dipping a hot blend of the mixed polysaccharide solution and genipin in KCl solution. The release of β-carotene was performed in simulated gastrointestinal condition at 37°C under the pH change method. Swelling characteristics, network stability, and release properties depended on the cross-linker concentration (Muhamad et al., 2011).

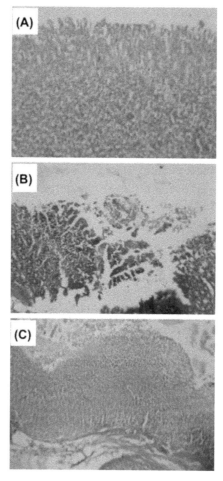

FIGURE 5.2   (See color insert.) Stomach histopathology of control rats (A), rats treated with pure ketoprofen (B), and rats treated with interpenetrating network hydrogel beads (C). (Hematoxylin-Eosin, 100×.). It is reprinted from (Kulkarni et al., 2012), Copyright 2012, with permission from Elsevier.

The investigation of κ-carrageenan for applications in regenerative medicine and tissue engineering and requires evaluation of its biocompatibility with the surrounding tissue of hydrogel implants. The cytotoxicity of κ-carrageenan hydrogels was tested *in vitro* on an L929 mouse fibroblast cell line and human poly-morphonuclear neutrophils cells. The *in vivo* study was practiced on rats regarding the inflammatory and immune reaction on subcutaneous implantation (Popa et al., 2014). The metabolism of the cells was not significantly affected while the inflammatory response was weak.

An interesting material with κ-carrageenan has been investigated in order to improve the drawbacks of conventional ophthalmic administration such as poor bioavailability. The ability of κ-carrageenan for ionotropic gelation in the presence of K^+ and Ca^{2+} ions was used to make an *in situ* gelling matrices for the delivery of the model drug, hydroxypropyl methylcellulose (HPMC). The formed network extended the time that the drug remained in the corneal area (Li et al., 2017).

Patches based on carrageenan cubic hydrogels were tested on porcine skin for the delivery of 5-aminolevulinic acid (Valenta et al., 2005). κ-Carrageenan hydrogels with embedded magnetic nanoparticles demonstrated the ability to instructively affect chondrogenic phenotype in human stem cells derived by adipose. The positive influence of the externally triggered biological processes applies to cell viability, cell content, and metabolic activity. This was a hopeful result for tissue engineering of cartilage (Popa et al., 2016).

Chemical networks originating from blends of κ-carrageenan and polyvinyl alcohol (PVA) were cross-linked by genipin. The active material β-carotene was dispersed in ethanol and mixed with the polymer blend. The subsequent cross-linking was used to immobilize β-carotene within the network matrix. Genipin enhanced the physical stability of the hydrogels, eliminated the burst release of the drug, and controlled its diffusion (Hezaveh and Muhamad, 2013). In an earlier study, the influence of κ-carrageenan incorporated in agarose gels on the diffusion of drugs was investigated. Very interestingly, the hydrophobic part of the amphiphilic test drugs was varied. The diffusion coefficient of all six drugs was apparently the same in agarose gels while it reduced in the polysaccharide-containing gels. In experiments of simultaneous diffusion of two drugs, the diffusion coefficient decreased to a much larger extent for the more hydrophobic drug in comparison to the less hydrophobic one. The authors put forward the possibility of exploiting hydrophobic cooperative effects in dual drug systems for controlled release in bi-phasic administration patterns (Sjöberg et al., 1999).

5.11 BIONANOCOMPOSITE HYDROGELS

Incorporation of inorganic (metallic and magnetic nanoparticles) in carrageenan networks may provide improved mechanical properties, stimuli responsiveness to external triggers and antimicrobial activity. Iron oxide nanofillers blended in κ-carrageenan gels induced changes in the swelling and elastic characteristics. Gelation in this system was achieved by K^+ ions. Although the network viscoelasticity was reinforced by the presence of nanoparticles, their charge tended to increase the swelling ratio. The influence of the presence of nanoparticles on the kinetics of release was tested with the model drug methylene blue. Its release was measured by UV-Vis in solutions containing lyophilized gel disks. Drug diffusion was coupled to polymer relaxation, and it was influenced by both the swelling and mechanical reinforcement caused by the iron oxide nanoparticles (Daniel Da Silva et al., 2012). Gold nanoparticles were embedded in a similarly cross-linked carrageenan hydrogel. The release kinetics and underlying mechanism of methylene blue could be tuned by the shape of the nanoparticles, i.e., spherical, short rod, or long rod (Salgueiro et al., 2013). Nanocomposite hydrogels with metallic nanoparticles have also been suggested for minimization of the release of methylene blue in the stomach and targeted release in the GIT (Hezaveh and Muhamad, 2012).

Nanocomposite hydrogel beads containing iron nanoparticles have been formed by in situ precipitation of iron ions in mixed κ-carrageenan and CS (Figure 5.3). Ionic cross-linking occurs between K^+ or Ca^{2+} ions and charged groups of the polysaccharides. Diclofenac sodium release from the beads could be adjusted by both the pH and the external alternative magnetic field, which enhanced the amount of released drug (Mahdavinia et al., 2015). Triple-stimuli responsive hydrogels of κ-carrageenan cross-linked by polyacrylic acid were obtained using paramagnetic iron oxide nanoparticles. Their swelling ratio was determined by temperature, pH, and external magnetic field, and these triggers defined *in vitro* release of deferasirox (Bardajee et al., 2013).

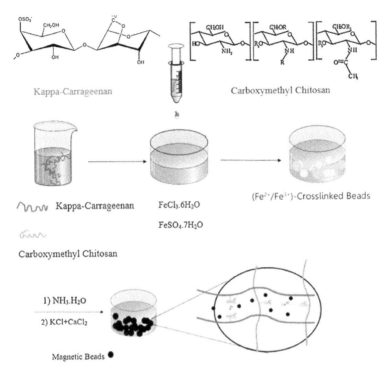

FIGURE 5.3 **(See color insert.)** Schematic diagram of the synthesis of magnetic κ-carrageenan/chitosan beads.
Reprinted from (Mahdavinia et al., 2015), Copyright 2015, with permission from Elsevier.

Antibacterial activity against E. coli and Bacillus was reported from ι-carrageenan/silver (Ag^0) nanocomposite hydrogels. Nanocolloidal silver was produced by reduction of $AgNO_3$ within the pre-synthesized hydrogels (Jayaramudu et al., 2013). Maghemite nanoparticles stabilized by electrostatic interactions in ι-carrageenan hydrogels showed biocompatibility and were able to induce apoptosis to human colon cancer cell line. With this system, magnetically driven targeted drug delivery for cancer therapy was proposed (Raman et al., 2015).

5.12 TISSUE ENGINEERING

The 3D mechanically stable structure of hydrogels, their ability for high water uptake and the consequent permeability to oxygen and nutrients make them unique for scaffolds for cell culture. This way, they mimic the extracellular matrix and are ideal for tissue engineering applications. Tissue regeneration was triggered in combinations of scaffolds with cells growth factors and genetic material. Tissue formation can be activated by the physicochemical and 3D spatial network properties (Barrère et al., 2008). Hydrogels from κ-carrageenan were proposed for tissue engineering as they were found to preserve the viability of incorporated fibroblast cells. The photocrosslinking reaction used offered versatility in elastic moduli, swelling ratio, and pore size distribution (Mihaila et al., 2013).

Vessel-like networks consisting of endothelial cells are considered essential for creating artificial tissues and organs (Mihaila et al., 2014). Vasculature is necessary to sustain nutrients and oxygen supply to thick artificial tissues. To this end, micro-sized hydrogel fibers with the potential for vascularized bone tissue engineering were developed. Fibers were made of κ-carrageenan by ionotropic gelation (K^+) and the wet spinning technique. They were subsequently coated with CS via electrostatic complexation for stability improvement. The fibers showed potential as a cell delivery system in 3D tissue engineering and as building blocks for the 3D cell containing hydrogels. Phenotype and in vitro functionality of osteoblast-like encapsulated cells was maintained after 21 days of culturing. After 21 days of encapsulation, cells were retrieved from the fibers (Mihaila et al., 2014). They could form endothelial-like colonies (Figure 5.4A) and tubular-like structures (Figure 5.4B).

Biocompatible cryogels made of carrageenan and gelatin were obtained by two different cross-linking agents (GA and EDC-NHS). Fibroblast cells (Cos-7) were cultured within the matrices, and their proliferation was evaluated (Sharma et al., 2013). In other hydrogels made of carrageenan, the potential for cartilage tissue engineering was addressed. Stem cells and growth factors were encapsulated, and the initial steps of chondrogenic differentiation were observed (Rocha et al., 2011). Syneresis of κ-carrageenan in blends with a C2-symmetric benzene-based supramolecular hydrogelator was reported. In the formed interpenetrating networks, cell growth and proliferation was achieved (Zhang et al., 2016a).

Polyelectrolyte complexes between CS and κ-carrageenan were reported to form porous scaffolds without the need of cross-linking. The porous precipitated complexes were stable up to physiological pH. Their mechanical properties were improved by increasing the amount of CS (Araujo et al., 2014). Carrageenan used as a surface modifier of graphene oxide was utilized to assist the binding of calcium ions and hence nucleation of hydroxyapatite. This functionalization led to biocompatible nano-composites. Cell mineralization was high on the modified surfaces. This system was suggested as a biomimetic material engineering of bone tissue (Liu et al., 2014).

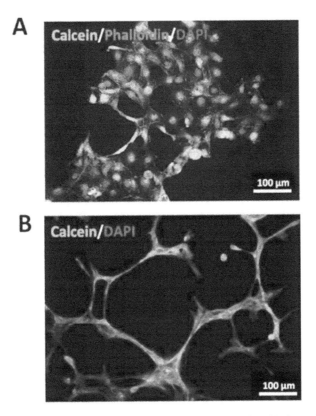

FIGURE 5.4 **(See color insert.)** Cells that were retrieved after 21 days of culture in fibers. DAPI staining is used for DNA and phalloidin for F-actin. Green-calcein represents live cells.
Adapted with permission from (Mihaila et al., 2014), Copyright 2014 American Chemical Society.

5.13　COMPLEXATION WITH MACROMOLECULAR STRUCTURES

The interactions between polyelectrolytes and charged drugs may be used for controlled loading and release. The complexation of the cationic drug doxazosin with the three types of carrageenan (ι, κ, and λ) in the form of tablets was studied with a doxazosin mesylate selective membrane electrode. The binding of doxazosin on carrageenan was found to occur with a double mechanism. Strong electrostatic interaction was responsible for initial binding while amphiphilicity of the drug-induced cooperative accumulation leading to charge aggregated structures. The linear charge density (sulfate groups), which depends on the polysaccharide type determined binding. The mechanisms of drug release were related to swelling and corrosion of the tablets and subsequent diffusion. Hydrophobic and electrostatic interactions also influenced drug release (Pavli et al., 2011). In an earlier study, the electrostatic binding of several drugs with different hydrophobicities had been reported, and the role of hydrophobic cooperativity had been highlighted. Additionally, the complexation of the drugs with κ-carrageenan in the helical or coiled conformation was tested. In 0.03 M NaCl κ-carrageenan chains are in the coil conformation while in 0.03 M KCl they form helices. In a helix, charge-charge distances along the macromolecular backbone are shorter in comparison to the coil. Consequently, it was found that drug complexation was higher when carrageenan was in the helix form (Caram-Lelham and Sundelöf, 1996).

Microspheres of carrageenan/gelatin complexes in water can be obtained by standard emulsion methodology. The size of the microspheres can be tuned by the surfactant and polymer content. Carrageenan of type I (mostly κ- and lesser λ- carrageenan) and gelatin type A were cross-linked with genipin to form spheres 500 μm in size. Washed and freeze-dried microparticles were loaded with isoniazid by dipping in drug solutions. Drug release rates were correlated to the cross-linking density (Devi and Maji, 2010). Nanoemulsions modified by polysaccharides for the topical delivery of ibuprofen were tested *in vitro* on rat skin. Drug permeation was increased by modification (Salim et al., 2012).

In a study where chondroitin sulfate was also tested, λ-carrageenan was coprecipitated with ciprofloxacin due to electrostatic interaction. Mucoadhesive properties of the microparticles were enhanced by the addition of Carbopol as it was evaluated by an "inclined plane" method on gastric mucin substrates. The produced formulations were considered promising for ocular drug administration, especially the ones originating from chondroitin sulfate (Gavini et al., 2016). In doubly coated alginate microparticles λ-carrageenan was used as an outer stabilizing layer in stomach conditions. The inner layer of CS assisted in the control of 5-aminosalicylic acid release (Karewicz et al., 2011).

Ionic complexation of carrageenan with CS is a way to produce well-defined nanoparticles for drug and protein delivery. Loading and controlled release capabilities have been tested by ovalbumin and extensively sustained release was observed. Such formulations are also low in toxicity and biocompatible (Grenha et al., 2010). CS/κ-carrageenan nanoparticles (about 200 nm in size) could be controlled by tripolyphosphate cross-linking (Rodrigues et al., 2012). Nanoparticles of larger size were obtained by including ι-carrageenan in the dual system. They were optimized for oral delivery of insulin with the low and high release in gastric and intestinal conditions, respectively (Sahoo et al., 2017). Carboxymethylated κ-carrageenan microparticles functionalized by lectin have also shown the potential for oral drug delivery of insulin (Leong et al., 2011).

Complexation of carrageenans with protamine is another example of polyelectrolyte complexation that leads to well-defined self-assembled nanoparticles with size and surface charge, which can be tuned by divalent salts (Dul et al., 2015).

Thermally induced gelation in reverse microemulsions has been found to produce κ-carrageenan nanogels with sizes smaller than 100 nm. The reversible thermoresponsive volume transitions of the nanogels could tune the release of methylene blue (Daniel Da Silva et al., 2011). Oil-in-water nanoemulsions stabilized by the cationic surfactant dodecyltrimethylammonium chloride could be modified by adsorption of κ-carrageenan layers and offered a system with possibilities in drug delivery of hydrophobic and/or charged drugs (Rosas-Durazo et al., 2011).

5.14 PROTEIN-LOADED NANOPARTICLES

Oppositely charged polysaccharides and proteins are customarily used for self-assembled micro- and nanoformulations exploiting the attractive electrostatic interactions (Turgeon et al., 2007; Schmitt and Turgeon, 2011; Grenha, 2012). Polysaccharide nanoparticles are biocompatible and biodegradable entities that can be used in protein, gene, and drug delivery applications (Nitta and Numata, 2013). Recently carrageenan has been reported as a polysaccharide with limited application in drug delivery (Dionísio et al., 2013). κ-Carrageenan in the sodium salt form was complexed with aminated pullulan in excess to produce positively charged nanoparticles ~190 nm in diameter (Dionísio et al., 2013). The obtained spherical nanoparticles had the ability to incorporate bovine serum albumin (BSA). A burst release of the protein that remained for 24 h was observed in phosphate buffer saline pH 7.4. Freeze drying in the presence of cryopectants did not interfere with nanoparticle properties like size and surface charge while the nanoparticles did not cause toxicity during *in vitro* tests on a Calu-3 cell line.

Curcumin-loaded nanoparticles have been produced based on Gantrez polymer in order to exploit the therapeutic properties of curcumin against hepatic diseases (D'Souza et al., 2013). κ-Carrageenan was one of the three polysaccharides introduced to create a series of different nanocarriers. Asialoglycoprotein receptor is overexpressed in hepatic disease and has a strong affinity for galactose-based carbohydrates. The size of the particles was in the order of 500 nm, and the ligand binding was proved by changes in their surface charge. Hepatic accumulation of carrageenan nanoparticles was intermediate in comparison to the other two polysaccharides as shown by in vivo biodistribution in rats and radiolabeled complexes (D'Souza et al., 2013).

5.15   CONCLUDING REMARKS

In this chapter, the pharmaceutical applications of carrageenan were presented. The role of this polysaccharide in prolonged drug release, experimental inflammation, and anti-HIV activity were illustrated. Several methodologies include carrageenan in drug stimuli-responsive hydrogels, bionanocomposites, and nanoparticles with potential for nanodelivery, tissue engineering and wound healing. The demonstrated results prove carrageenan's importance as a highly charged natural polymer with bioadhesive properties that is able to interact with proteins, drugs, and tissues and also form the basis for buccal films. The progress in the use of carrageenan in pharmaceutical applications during the last decades will surely allow this versatile polysaccharide to take part in the developments in biomedicine and biotechnology of the near future.

KEYWORDS

- **anti-HIV activity**
- **bionanocomposites**
- **carrageenan**
- **hydrogels**
- **tissue engineering**
- **wound healing**

REFERENCES

Abdelazeem, A. H., et al., (2014). Novel pyrazolopyrimidine derivatives targeting COXs and iNOS enzymes, design, synthesis and biological evaluation as potential anti-inflammatory agents. *European Journal of Pharmaceutical Sciences, 62*, 197–211.

Almasirad, A., et al., (2014). Synthesis, analgesic and anti-inflammatory activities of new methyl-imidazolyl-1,3,4-oxadiazoles and 1,2,4-triazoles. *DARU Journal of Pharmaceutical Sciences, 22*(1), 22.

Alvarez-Lorenzo, C., et al., (2013). Crosslinked ionic polysaccharides for stimuli-sensitive drug delivery. *Advanced Drug Delivery Reviews, 65*(9), 1148–1171.

Anderson, W., et al., (1965). The anticoagulant activity of carrageenan. *Journal of Pharmacy and Pharmacology, 17*(10), 647–654.

Araujo, J. V., et al., (2014). Novel porous scaffolds of pH-responsive chitosan/carrageenan-based polyelectrolyte complexes for tissue engineering. *Journal of Biomedical Materials Research Part A., 102*(12), 4415–4426.

Baba, M., et al., (1988). Sulfated polysaccharides are potent and selective inhibitors of various enveloped viruses, including herpes simplex virus, cytomegalovirus, vesicular stomatitis virus, and human immunodeficiency virus. *Antimicrobial Agents and Chemotherapy, 32*(11), 1742–1745.

Bani-Jaber, A., & Al-Ghazawi, M., (2005). Sustained release characteristics of tablets prepared with mixed matrix of sodium carrageenan and chitosan: Effect of polymer weight ratio, dissolution medium, and drug type. *Drug Development and Industrial Pharmacy, 31*(3), 241–247.

Bardajee, G. R., et al., (2013). Kappa carrageenan-g-poly (acrylic acid)/SPION nanocomposite as a novel stimuli-sensitive drug delivery system. *Colloid and Polymer Science, 291*(12), 2791–2803.

Barrère, F., et al., (2008). Advanced biomaterials for skeletal tissue regeneration: Instructive and smart functions. *Materials Science and Engineering: R: Reports, 59*(1), 38–71.

Basit, A. W., et al., (1999). Formulation of ranitidine pellets by extrusion-spheronization with little or no microcrystalline cellulose. *Pharmaceutical Development and Technology, 4*(4), 499–505.

Bertram, U., & Bodmeier, R., (2012). Effect of polymer molecular weight and of polymer blends on the properties of rapidly gelling nasal inserts. *Drug Development and Industrial Pharmacy, 38*(6), 659–669.

Bertram, U., et al., (2010). *In situ* gelling nasal inserts for influenza vaccine delivery. *Drug Development and Industrial Pharmacy, 36*(5), 581–593.

Biruss, B., & Valenta, C., (2006). Skin permeation of different steroid hormones from polymeric coated liposomal formulations. *European Journal of Pharmaceutics and Biopharmaceutics, 62*(2), 210–219.

Boxenbaum, H. G., & Dairman, W., (1977). Evaluation of an animal model for the screening of compounds potentially useful in human ulcerative colitis: Effect of salicylazosulfapyridine and prednisolone on carrageenan-induced ulceration of the large intestine of the guinea pig. *Drug Development and Industrial Pharmacy, 3*(2), 121–130.

Caram-Lelham, N., & Sundelöf, L. O., (1996). The effect of hydrophobic character of drugs and helix-coil transition of κ-carrageenan on the polyelectrolyte-drug interaction. *Pharmaceutical Research, 13*(6), 920–925.

Caramella, C. M., et al., (2015). Mucoadhesive and thermogelling systems for vaginal drug delivery. *Advanced Drug Delivery Reviews, 92*(Supplement C), 39–52.

D'Souza, A. A., et al., (2013). Comparative *in silico–in vivo* evaluation of ASGP-R ligands for hepatic targeting of curcumin gantrez nanoparticles. *The AAPS Journal, 15*(3), 696–706.

Daniel Da Silva, A. L., et al., (2012). Impact of magnetic nanofillers in the swelling and release properties of κ-carrageenan hydrogel nanocomposites. *Carbohydrate Polymers, 87*(1), 328–335.

Daniel-Da-Silva, A. L., et al., (2011). Synthesis and swelling behavior of temperature responsive κ-carrageenan nanogels. *Journal of Colloid and Interface Science, 355*(2), 512–517.

De Silva, D. A., et al., (2011). Development of a PVP/kappa-carrageenan/PEG hydrogel dressing for wound healing applications in Sri Lanka. *Journal of the National Science Foundation of Sri Lanka, 39*(1), 25–33.

Deghrigue, M., et al., (2014). Pharmacological evaluation of the semi-purified fractions from the soft coral Eunicella singularis and isolation of pure compounds. *DARU Journal of Pharmaceutical Sciences, 22*(1), 64.

Devi, N., & Maji, T. K., (2010). Microencapsulation of isoniazid in genipin-crosslinked gelatin-A-κ-carrageenan polyelectrolyte complex. *Drug Development and Industrial Pharmacy, 36*(1), 56–63.

Dionísio, M., et al., (2013). Pullulan-based nanoparticles as carriers for transmucosal protein delivery. *European Journal of Pharmaceutical Sciences, 50*(1), 102–113.

Djupesland, P. G., (2013). Nasal drug delivery devices: Characteristics and performance in a clinical perspective—a review. *Drug Delivery and Translational Research, 3*(1), 42–62.

Dul, M., et al., (2015). Self-assembled carrageenan/protamine polyelectrolyte nanoplexes— Investigation of critical parameters governing their formation and characteristics. *Carbohydrate Polymers, 123*, 339–349.

Eccles, R., et al., (2015). Efficacy and safety of iota-carrageenan nasal spray versus placebo in early treatment of the common cold in adults: The ICICC trial. *Respiratory Research, 16*, 121.

Fernández-Romero, J. A., et al., (2007). Carrageenan/MIV-150 (PC-815), a combination microbicide. *Sexually Transmitted Diseases, 34*(1), 9–14.

Fernández-Romero, J. A., et al., (2012). Zinc acetate/carrageenan gels exhibit potent activity *in vivo* against high-dose herpes simplex virus 2 vaginal and rectal challenge. *Antimicrobial Agents and Chemotherapy, 56*(1), 358–368.

Gavini, E., et al., (2016). Engineered microparticles based on drug-polymer coprecipitates for ocular-controlled delivery of ciprofloxacin: Influence of technological parameters. *Drug Development and Industrial Pharmacy, 42*(4), 554–562.

Grassauer, A., et al., (2008). Iota-carrageenan is a potent inhibitor of rhinovirus infection. *Virology Journal, 5*(1), 107.

Grenha, A., (2012). Chitosan nanoparticles: A survey of preparation methods. *Journal of Drug Targeting, 20*(4), 291–300.

Grenha, A., et al., (2010). Development of new chitosan/carrageenan nanoparticles for drug delivery applications. *Journal of Biomedical Materials Research Part A., 92A*(4), 1265–1272.

Gu, J., et al., (2015). Biodegradable film for the targeted delivery of siRNA-loaded nanoparticles to vaginal immune cells. *Molecular Pharmaceutics, 12*(8), 2889–2903.

Hadj, A. H., et al., (2015). Physico-chemical characterization and pharmacological evaluation of sulfated polysaccharides from three species of Mediterranean brown algae of the genus cystoseira. *DARU Journal of Pharmaceutical Sciences, 23*(1), 1.

Hanawa, T., et al., (2004). Development of patient-friendly preparations: Preparation of a new allopurinol mouthwash containing polyethylene(oxide) and carrageenan. *Drug Development and Industrial Pharmacy, 30*(2), 151–161.

Hariharan, M., et al., (1997). Optimization of sustained-release tablet formulations: A four-component mixture experiment. *Pharmaceutical Development and Technology, 2*(4), 365–371.

Hezaveh, H., & Muhamad, I. I., (2012). The effect of nanoparticles on gastrointestinal release from modified κ-carrageenan nanocomposite hydrogels. *Carbohydrate Polymers, 89*(1), 138–145.

Hezaveh, H., & Muhamad, I. I., (2013). Controlled drug release via minimization of burst release in pH-response kappa-carrageenan/polyvinyl alcohol hydrogels. *Chemical Engineering Research and Design, 91*(3), 508–519.

Hosseinzadeh, H., (2010). Controlled release of diclofenac sodium from pH-responsive carrageenan-g-poly(acrylic acid) superabsorbent hydrogel. *Journal of Chemical Sciences, 122*(4), 651–659.

Jayaramudu, T., et al., (2013). Iota-carrageenan-based biodegradable Ag0 nanocomposite hydrogels for the inactivation of bacteria. *Carbohydrate Polymers, 95*(1), 188–194.

Karewicz, A., et al., (2011). New bilayer-coated microbead system for controlled release of 5-aminosalicylic acid. *Polymer Bulletin, 66*(3), 433–443.

Kawabata, A., et al., (1992). l-Leucyl-l-arginine, naltrindole and d-arginine block antinociception elicited by l-arginine in mice with carrageenin-induced hyperalgesia. *British Journal of Pharmacology, 107*(4), 1096–1101.

Keppeler, S., et al., (2009). Cross-linked carrageenan beads for controlled release delivery systems. *Carbohydrate Polymers, 78*(4), 973–977.

Khan, A. K., et al., (2017). Carrageenan based bionanocomposites as drug delivery tool with special emphasis on the influence of ferromagnetic nanoparticles. *Oxidative Medicine and Cellular Longevity*, 8158315.

Khedr, M. A., et al., (2014). Repositioning of 2,4-dichlorophenoxy acetic acid as a potential anti-inflammatory agent: *In silico* and pharmaceutical formulation study. *European Journal of Pharmaceutical Sciences, 65*, 130–138.

Kianfar, F., et al., (2012). Novel films for drug delivery via the buccal mucosa using model soluble and insoluble drugs. *Drug Development and Industrial Pharmacy, 38*(10), 1207–1220.

Kianfar, F., et al., (2014). Development and physicomechanical characterization of carrageenan and poloxamer-based lyophilized matrix as a potential buccal drug delivery system. *Drug Development and Industrial Pharmacy, 40*(3), 361–369.

Koenighofer, M., et al., (2014). Carrageenan nasal spray in virus confirmed common cold: Individual patient data analysis of two randomized controlled trials. *Multidisciplinary Respiratory Medicine, 9*(1), 57.

Kranz, H., et al., (2009). Drug release from MCC- and carrageenan-based pellets: Experiment and theory. *European Journal of Pharmaceutics and Biopharmaceutics, 73*(2), 302–309.

Kulkarni, R. V., et al., (2012). pH-responsive interpenetrating network hydrogel beads of poly(acrylamide)-g-carrageenan and sodium alginate for intestinal targeted drug delivery: Synthesis, *in vitro* and *in vivo* evaluation. *Journal of Colloid and Interface Science, 367*(1), 509–517.

Leibbrandt, A., et al., (2010). Iota-carrageenan is a potent inhibitor of influenza A virus infection. *PLOS ONE, 5*(12), e14320.

Leong, K. H., et al., (2011). Lectin-functionalized carboxymethylated kappa-carrageenan microparticles for oral insulin delivery. *Carbohydrate Polymers, 86*(2), 555–565.

Levendosky, K., et al., (2015). Griffithsin and carrageenan combination to target herpes simplex virus 2 and human papillomavirus. *Antimicrobial Agents and Chemotherapy, 59*(12), 7290–7298.

Li, L., et al., (2014). Carrageenan and its applications in drug delivery. *Carbohydrate Polymers, 103*, 1–11.

Li, P., et al., (2017). A novel ion-activated in situ gelling ophthalmic delivery system based on κ-carrageenan for acyclovir. *Drug Development and Industrial Pharmacy*, 1–8.

Ling, G., et al., (2016). Nanostructured lipid–carrageenan hybrid carriers (NLCCs) for controlled delivery of mitoxantrone hydrochloride to enhance anticancer activity bypassing the BCRP-mediated efflux. *Drug Development and Industrial Pharmacy, 42*(8), 1351–1359.

Liu, H., et al., (2014). Biomimetic and cell-mediated mineralization of hydroxyapatite by carrageenan functionalized graphene oxide. *ACS Applied Materials & Interfaces, 6*(5), 3132–3140.

Liu, Y., et al., (2009). Effect of carrageenan on poloxamer-based in situ gel for vaginal use: Improved *in vitro* and *in vivo* sustained-release properties. *European Journal of Pharmaceutical Sciences, 37*(3), 306–312.

Ludwig, M., et al., (2013). Efficacy of a carrageenan nasal spray in patients with common cold: A randomized controlled trial. *Respiratory Research, 14*(1), 124.

Maguire, R. A., et al., (1998). Carrageenan-based nonoxynol-9 spermicides for prevention of sexually transmitted infections. *Sexually Transmitted Diseases, 25*(9), 494–500.

Mahdavinia, G. R., et al., (2015). Magnetic/pH-responsive beads based on carboxymethyl chitosan and κ-carrageenan and controlled drug release. *Carbohydrate Polymers, 128*, 112–121.

Malafaya, P. B., et al., (2007). Natural–origin polymers as carriers and scaffolds for biomolecules and cell delivery in tissue engineering applications. *Advanced Drug Delivery Reviews, 59*(4), 207–233.

Mihaila, S. M., et al., (2013). Photocrosslinkable kappa-carrageenan hydrogels for tissue engineering applications. *Advanced Healthcare Materials, 2*(6), 895–907.

Mihaila, S. M., et al., (2014). Fabrication of endothelial cell-laden carrageenan microfibers for microvascularized bone tissue engineering applications. *Biomacromolecules, 15*(8), 2849–2860.

Moyana, T. N., & Lalonde, J. M., (1990). Carrageenan-induced intestinal injury in the rat—a model for inflammatory bowel disease. *Ann. Clin. Lab. Sci., 20*(6), 420–426.

Muhamad, I. I., et al., (2011). Genipin-cross-linked kappa-carrageenan/carboxymethyl cellulose beads and effects on beta-carotene release. *Carbohydrate Polymers, 83*(3), 1207–1212.

Mundargi, R. C., et al., (2007). Development of polysaccharide-based colon targeted drug delivery systems for the treatment of amoebiasis. *Drug Development and Industrial Pharmacy, 33*(3), 255–264.

Nair, A. V., et al., (2016). Cyclic [small beta]-(1[rightward arrow]3) (1[rightward arrow]6) glucan/carrageenan hydrogels for wound healing applications. *RSC Advances, 6*(100), 98545–98553.

Netto, M. V., et al., (2005). Wound healing in the cornea: A review of refractive surgery complications and new prospects for therapy. *Cornea, 24*(5), 509–522.

Nitta, S. K., & Numata, K., (2013). Biopolymer-based nanoparticles for drug/gene delivery and tissue engineering. *International Journal of Molecular Sciences, 14*(1), 1629–1654.

Pavli, M., et al., (2011). Doxazosin–carrageenan interactions: A novel approach for studying drug-polymer interactions and relation to controlled drug release. *International Journal of Pharmaceutics, 421*(1), 110–119.

Picker, K. M., (1999a). Matrix tablets of carrageenans. I: A compaction study. *Drug Development and Industrial Pharmacy, 25*(3), 329–337.

Picker, K. M., (1999b). Matrix tablets of carrageenans. II: Release behavior and effect of added cations. *Drug Development and Industrial Pharmacy, 25*(3), 339–346.

Piyakulawat, P., et al., (2007). Preparation and evaluation of chitosan/carrageenan beads for controlled release of sodium diclofenac. *AAPS Pharm. Sci. Tech., 8*(4), 120.

Popa, E. G., et al., (2014). Evaluation of the *in vitro* and *in vivo* biocompatibility of carrageenan-based hydrogels. *Journal of Biomedical Materials Research Part A., 102*(11), 4087–4097.

Popa, E. G., et al., (2016). Magnetically-responsive hydrogels for modulation of chondrogenic commitment of human adipose-derived stem cells. *Polymers, 8*(2), 28.

Raman, M., et al., (2015). Biocompatible ι-carrageenan-γ-maghemite nanocomposite for biomedical applications–synthesis, characterization and *in vitro* anticancer efficacy. *Journal of Nanobiotechnology, 13*, 18.

Raveendran, S., et al., (2013). Pharmaceutically versatile sulfated polysaccharide based bionano platforms. *Nanomedicine: Nanotechnology, Biology and Medicine, 9*(5), 605–626.

Roberts, J. N., et al., (2011). Effect of Pap smear collection and carrageenan on cervicovaginal human papillomavirus-16 infection in a rhesus macaque model. *JNCI Journal of the National Cancer Institute, 103*(9), 737–743.

Rocha, P. M., et al., (2011). Encapsulation of adipose-derived stem cells and transforming growth factor-β1 in carrageenan-based hydrogels for cartilage tissue engineering. *Journal of Bioactive and Compatible Polymers, 26*(5), 493–507.

Rodrigues, S., et al., (2012). Chitosan/carrageenan nanoparticles: Effect of cross-linking with tripolyphosphate and charge ratios. *Carbohydrate Polymers, 89*(1), 282–289.

Rosario, N. L., & Ghaly, E. S., (2002). Matrices of water-soluble drug using natural polymer and direct compression method. *Drug Development and Industrial Pharmacy, 28*(8), 975–988.

Rosas-Durazo, A., et al., (2011). Development and characterization of nanocapsules comprising dodecyltrimethylammonium chloride and κ-carrageenan. *Colloids and Surfaces B: Biointerfaces, 86*(1), 242–246.

Rupenthal, I. D., et al., (2011). Ion-activated *in situ* gelling systems for antisense oligo-deoxynucleotide delivery to the ocular surface. *Mol. Pharmaceutics, 8*(6), 2282–2290.

Saha, D., & Bhattacharya, S., (2010). Hydrocolloids as thickening and gelling agents in food: A critical review. *Journal of Food Science and Technology, 47*(6), 587–597.

Sahoo, P., et al., (2017). Optimization of pH-responsive carboxymethylated iota-carrageenan/chitosan nanoparticles for oral insulin delivery using response surface methodology. *Reactive and Functional Polymers, 119*, 145–155.

Salgueiro, A. M., et al., (2013). κ-Carrageenan hydrogel nanocomposites with release behavior mediated by morphological distinct Au nanofillers. *Carbohydrate Polymers, 91*(1), 100–109.

Salim, N., et al., (2012). Modification of palm kernel oil esters nanoemulsions with hydrocolloid gum for enhanced topical delivery of ibuprofen. *International Journal of Nanomedicine, 7*, 4739–4747.

Santin, J. R., et al., (2013). Chemical synthesis, docking studies and biological effects of a pan peroxisome proliferator-activated receptor agonist and cyclooxygenase inhibitor. *European Journal of Pharmaceutical Sciences, 48*(4), 689–697.

Schaeffer, D. J., & Krylov, V. S., (2000). Anti-HIV activity of extracts and compounds from algae and cyanobacteria. *Ecotoxicology and Environmental Safety, 45*(3), 208–227.

Schmitt, C., & Turgeon, S. L., (2011). Protein/polysaccharide complexes and coacervates in food systems. *Advances in Colloid and Interface Science, 167*(1), 63–70.

Šentjurc, M., et al., (1999). Liposomes as a topical delivery system: The role of size on transport studied by the EPR imaging method. *Journal of Controlled Release, 59*(1), 87–97.

Sharma, A., et al., (2013). Three-dimensional supermacroporous carrageenan-gelatin cryogel matrix for tissue engineering applications. *BioMed. Research International, 2013*, 15.

Silva, D. A. D., et al., (2011). Development of a PVP/kappa-carrageenan/PEG hydrogel dressing for wound healing applications in Sri Lanka. *Journal of the National Science Foundation of Sri Lanka, 39*(1), 25–33.

Singer, R., et al., (2011). The nonnucleoside reverse transcriptase inhibitor MIV-150 in carrageenan gel prevents rectal transmission of simian/human immunodeficiency virus infection in macaques. *Journal of Virology, 85*(11), 5504–5512.

Sjöberg, H., et al., (1999). How interactions between drugs and agarose-carrageenan hydrogels influence the simultaneous transport of drugs. *Journal of Controlled Release, 59*(3), 391–400.

Skoler-Karpoff, S., et al. (2008). Efficacy of carraguard for prevention of HIV infection in women in South Africa: A randomized, double-blind, placebo-controlled trial. *The Lancet, 372*(9654), 1977–1987.

Song, K. H., & Eddington, N. D., (2012). The influence of AT1002 on the nasal absorption of molecular weight markers and therapeutic agents when co-administered with bioadhesive polymers and an AT1002 antagonist, AT1001. *Journal of Pharmacy and Pharmacology, 64*(1), 30–39.

Soraya, H., et al., (2015). Angiogenic effect of the aqueous extract of Cynodon dactylon on human umbilical vein endothelial cells and granulation tissue in rat. *DARU Journal of Pharmaceutical Sciences, 23*(1), 10.

Sosnik, A., et al., (2014). Mucoadhesive polymers in the design of nano-drug delivery systems for administration by non-parenteral routes: A review. *Progress in Polymer Science, 39*(12), 2030–2075.

Teller, R. S., et al., (2014). Intravaginal flux controlled pump for sustained release of macromolecules. *Pharmaceutical Research, 31*(9), 2344–2353.

Thrimawithana, T. R., et al., (2011). Effect of cations on the microstructure and *in-vitro* drug release of κ- and ι-carrageenan liquid and semi-solid aqueous dispersions. *Journal of Pharmacy and Pharmacology, 63*(1), 11–18.

Turgeon, S. L., et al., (2007). Protein–polysaccharide complexes and coacervates. *Current Opinion in Colloid & Interface Science, 12*(4), 166–178.

Valenta, C., (2005). The use of mucoadhesive polymers in vaginal delivery. *Advanced Drug Delivery Reviews, 57*(11), 1692–1712.

Valenta, C., et al., (2005). Skin permeation and stability studies of 5-aminolevulinic acid in a new gel and patch preparation. *Journal of Controlled Release, 107*(3), 495–501.

Valle, B. L., et al., (2016). Use of κ-carrageenan, chitosan and carbopol 974P in extruded and spheronized pellets that are devoid of MCC. *Drug Development and Industrial Pharmacy, 42*(11), 1903–1916.

Vinegar, R., et al., (1973). Some quantitative temporal characteristics of carrageenin-induced pleurisy in the rat. *Proc. Soc. Exp. Biol. Med., 143*(3), 711–714.

Vlieghe, P., et al., (2002). Synthesis of new covalently bound κ-carrageenan–AZT conjugates with improved anti-HIV activities. *Journal of Medicinal Chemistry, 45*(6), 1275–1283.

Winter, C. A., et al., (1962). Carrageenin-induced edema in hind paw of the rat as an assay for anti-inflammatory drugs. *Proc. Soc. Exp. Biol. Med., 111*(3), 544–547.

Woodsong, C., & Holt, J. D. S., (2015). Acceptability and preferences for vaginal dosage forms intended for prevention of HIV or HIV and pregnancy. *Advanced Drug Delivery Reviews, 92*, 146–154.

Yoo, A., & Kleinebudde, P., (2009). Spheronization of small extrudates containing κ-carrageenan. *Journal of Pharmaceutical Sciences, 98*(10), 3776–3787.

Zhang, J., et al., (2016a). Tuning syneresis properties of kappa-carrageenan hydrogel by C2-symmetric benzene-based supramolecular gelators. *Macromolecular Chemistry and Physics, 217*(10), 1197–1204.

Zhang, L., et al., (2013). Incorporation of monodisperse oligoethyleneglycol amino acids into anticonvulsant analogues of galanin and neuropeptide Y provides peripherally acting analgesics. *Molecular Pharmaceutics, 10*(2), 574–585.

Zhang, S., et al., (2016b). Development of protocatechualdehyde proliposomes-based sustained-release pellets with improved bioavailability and desired pharmacokinetic behavior for angina chronotherapy. *European Journal of Pharmaceutical Sciences, 93,* 341–350.

CHAPTER 6

Pharmaceutical Application of Chitosan Derivatives

FIONA CONCY RODRIGUES, KRIZMA SINGH, and GOUTAM THAKUR

Department of Biomedical Engineering, Manipal Institute of Technology, Manipal Academy of Higher Education, Manipal – 576104, Karnataka, India

ABSTRACT

Chitosan (CS) is a well-known polysaccharide prepared by deacetylation of chitin. It is amongst the most abundant polysaccharides in the world and is known to be a versatile biopolymer due to its biocompatibility and biodegradability. Recent studies have highlighted its potent biological responses in pharmaceutical applications such as wound healing, tissue engineering, biomedical adhesives, etc. However, to modulate and enhance its biological activity, various chemical or enzymatic modifications are made to its structure to obtain its derivatives, which behave as better models. This chapter elaborates the pharmaceutical efficiency of various CS derivatives, which have been prepared by conventional chemical and enzymatic processes.

6.1 INTRODUCTION

Chitin and chitosan (CS) are promising classes of biopolymers and have been investigated in detail for their biomedical applications in the last few decades. Chitin is claimed to be the most abundantly occurring polysaccharide on the planet and is extracted from the exoskeletons of main crustaceans, insect cuticles, and fungi (Kaur, 2014). Chitin demonstrates insolubility in aqueous and common organic solvent sowing to its hydrophobic nature.

The most common method to acquire solubility in aqueous acidic medium is by converting chitin to CS through a deacetylation process. CS, a linear polymer, is composed of β-(1-4)-linked D-glucosamine and N-acetyl-D-glucosamine which is randomly distributed within the polymer. It has been observed that the majority of the existing polysaccharides exhibit either a neutral or negative charged when in an acidic environment. Interestingly, CS exhibits cationic nature, which enables it to form complexes or multilayer structures in conjugation with negatively charged synthetic or natural polymers. Due to its unique characteristics, owing to its natural origin such as nontoxicity, biocompatibility, and biodegradability, CS has been widely used for various pharmaceutical and biomedical applications as well as textile, paper, food industries, agriculture, and medicine (Kumar, Dutta, and Tripathi, 2004). Further, the presence of hydroxyl and the amino groups in the CS structure can be ameliorated. The structural modification of the native CS holds promise for new functional properties and broadening the array of applications.

The general aspects including the structure of chitin and CS, and the associated isolation processes are discussed in this chapter. Subsequently, the current state of knowledge of CS derivatives processes involved in the structural elucidation of the native CS molecule is focused. The last part of the chapter discusses about the recent advancements in the applications of CS in wound healing, tissue engineering, and drug delivery.

6.2   CHITOSAN (CS) STRUCTURE

6.2.1   STRUCTURES OF CHITIN AND CHITOSAN (CS)

Similar to cellulose, chitin exhibits similar functionality, as they are structural polysaccharides; however, they differ in their properties. Chitin is highly hydrophobic and insoluble in water and most organic solvents (Rinaudo, 2006). It is a structural biopolymer, which has a similar function to that of collagen in the higher animals and cellulose in plants. Plants produce cellulose in their cell walls, in similar way insects and crustaceans produce chitin in their shells as described by Zargar et al., (2015). Chitin can be obtained from the exoskeleton of arthropods or cell walls of yeast as it exists as an ordered crystalline microfibril, which plays an integral part of its structural composition (Raabe, Al-Sawalmih, and Yi, 2007). Chitin is a homopolymer of 2-acetamido-2-deoxy-bD-glucopyranose, although some of the glucopyranose residues are in the deacetylated form as 2-amino-2-deoxy-b-D-glucopyranose, whereas CS's majority of the glucopyranose residues are in the deacetylated form. Rarely has chitin been found in its N-acetyl or acetamido form, similarly chitosan is not always found in its deacetylated form except under rigorous conditions. Neither is found in the pure state in nature but in conjunction with other polysaccharides, proteins, and perhaps minerals (Zargar et al., 2015). Chitin may be considered as cellulose with the hydroxyl at position C2 substituted with an acetamido group. Both are polymers of structural monosaccharides consisting of β-(1-4)-2-acetamido-2-deoxy-b-D-glucose and β-(1-4)2-deoxy-b-D-glucopyranose units, respectively (Figure 6.1).

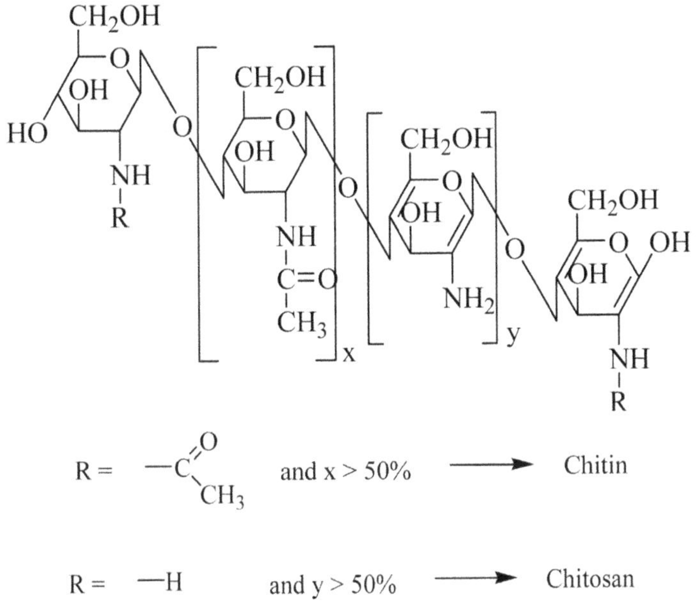

FIGURE 6.1 Structure of chitin and chitosan.
(Reproduced from E. Khor, Chitin: Fulfilling a Biomaterials Promise, Elsevier, Amsterdam 2001, with permission from Elsevier)

6.3 CHITOSAN (CS) ISOLATION

Chitosan is obtained from the raw material called chitin. The potential sources for the extraction of chitin as described in earlier section are the exoskeleton of crustaceans, mainly crabs and shrimps. Extraction of chitin from shrimps is relatively easier because of their thin shells. Additionally, there is no standard isolation process of chitin as there are different sources due to their diversities in nature and structure, and hence, they need different treatment. Conventionally, the isolation process involves the basic steps of demineralization, deproteinization, and decolorization, which can be achieved using either of the two treatments, i.e., chemical or biological (enzymatic treatment or fermentation) (Cheung et al., 2015).

6.3.1 CHEMICAL METHOD

In this treatment, the initial step involves the procuring of the shells from the different sources. These shells are then carefully cleaned and dried, followed by grinding it to small pieces or reducing it to a powder (Abdou and Nagy, 2007). Further, a demineralization process is carried out by treating it with strong acids such as hydrochloric acid, sulphuric acid, nitric acid, etc. This is followed by deproteination of the shells by an alkaline treatment using sodium hydroxide. Finally, the deacetylation process involves the treatment using sodium hydroxide (with higher concentration), with heating, to produce CS. The deacetylation process basically works by hydrolysis of the acetamide group as schematically represented in Figure 6.2. To obtain the final product, the resultant product is dissolved in an acidic medium (viz., diluted acetic acid, formic acid or lactic acid), followed by a final precipitation with dilute sodium hydroxide (Mini and Josef, 2011).

6.3.2 ENZYMATIC METHOD

This is an alternative method to produce CS from crustacean shells. Unlike chemical methods, demineralization of crustacean shells is carried out by the lactic acid-producing bacteria. The reaction works in a way that when the lactic acid reacts with calcium carbonate, it leads to the formation of calcium lactate. The calcium lactate is precipitated and removed (Imen and Fatih, 2016). Further, certain enzymes proteases which are extracted from bacteria are used for deproteination of crustacean shells. This is achieved by fermenting the crustacean with different bacterial species such as *Pseudomonas aeruginosa* K-187, *Serratia marcescens* FS-3, and *Bacillus subtilis* (Jo et al., 2008). Deacetylation of chitin is carried out by chitin deacetylase (Cai et al., 2006). The enzyme, chitin deacetylase, was first found in *Mucor rouxii* and later was produced from bacteria such as Serratia sp. and Bacillus sp. which were used to generate CS (Imen and Fatih, 2016).

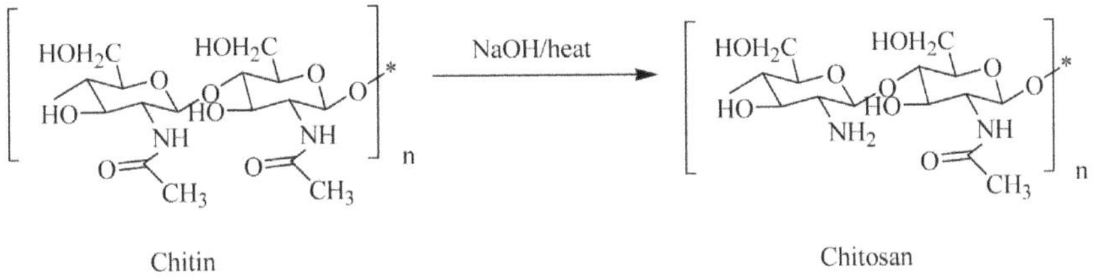

Chitin   Chitosan

FIGURE 6.2   Chitin to chitosan after deacetylation.
(Reproduced with modification from E. Khor, Chitin: Fulfilling a Biomaterials Promise, Elsevier, Amsterdam 2001, with permission from Elsevier).

6.4   CHITOSAN (CS) DERIVATIVES

Due to the innumerable unique biological traits associated with CS such as its biocompatibility, biodegradability, and nontoxicity, CS is a versatile biopolymer with varying applications in multiple fields (Agarwal et al., 2015). In the pharmaceutical domain particularly, CS has proved to be a potent biopolymer for drug delivery because of its mucoadhesion ability and permeation enhancement effects, which are highly desirable properties while designing targeted drug delivery systems (Cheung et al., 2015; Elieh-Ali-Komi and Hamblin, 2016; Kato, Onishi and Machida, 2003). However, despite all these advantages, there are certain properties associated with CS, which limits its usage. Some of the major disadvantages are its poor solubility at neutral pH, which limits its usage in certain delivery systems. Another limitation associated with CS in drug delivery aspects is its inability to have controlled release patterns because of the rapid absorption of water and a high swelling degree. This, in turn, leads to rapid drug release, which is undesirable while working with drug delivery systems (Prabaharan, 2008). To overcome these limitations, modifications are made to the native structure of CS to synthesize CS derivatives. The derivatives have a similar fundamental structure of CS but are tweaked at different reactive sites through chemical routes to obtain subsidiaries with enhanced properties. These modifications not only improve its solubility but also improve its sustainable release patterns, thereby widening its applications (Saikia and Gogoi, 2015).

CS indicates a relatively simple structure and can be easily modified. It has mainly three effective reactive sites for modification, mainly the primary hydroxy group, the secondary hydroxyl group, and the primary amine group, as indicated in Figure 6.3.

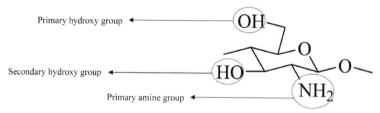

Primary hydroxy group

Secondary hydroxy group

Primary amine group

FIGURE 6.3   (See color insert.) Reactive sites for chitosan modifications.
Source: Agarwal et al., 2015, with permission from Journal of Pharma Research

Based on these sites of modifications, synthesis of a number of derivatives of CS has been carried out, and its applications have been investigated. Quaternary, thiolated, hydrophobic, grafted, and cross-linked CS are some of the main derivatives of CS that have indicated significant applications in the pharmaceutical industry and hence, have been discussed in this section (Figure 6.4).

6.4.1   QUATERNARY

Quaternary CS is a class of CS derivatives, which have been extensively studied. These derivatives improve CS's properties in a number of ways, such as its solubility at a wide range of pH and further stabilizing it at different pH by controlling its cationic character. It also leads to enhanced mucoadhesive and drug penetration properties, which are highly desirable properties when associated with the pharmaceutical domain (Ahmed and Aljaeid, 2016). One of the common ways of synthesizing quaternary CSs through the synthetic route is by varying the degree of substitution (DS), usually via methyl iodide or dimethylsulfate (Douglas, Rejane and Sergio, 2011). So far, the derivatives that have been developed are of the N-alkyl or quaternary ammonium. Most N-alkylated CS derivatives, (such as diethyl methyl, triethyl, trimethyl chitosans, etc.) are obtained by substituting the primary amine group with a suitable aldehyde or ketone. A general procedure to synthesize quarternary CSs would involve the constant stirring of CS with glycidyl trimethyl ammonium chloride (GTMAC) for 24 hours at 60°C (Spinelli, Laranjeira and Fávere, 2004). The usage of two specific quarternary derivatives- N-trimethyl chitosan (TMC) and N-(2-hydroxyl) propyl-3-trimethyl ammonium chitosan chloride (HTCC) have become common. The general method for the synthesis of TMC involves CS being dispersed in a solution of methyl iodide and methylpyrrolidone in the presence of NaOH at 60°C for 6 hours. It is followed by precipitation with acetone and further separation by centrifugation (Kean, Roth, and Thanou, 2005; Raphael et al., 2011; Sadeghi et al., 2008). Alternatively, HTCC is synthesized by substituting the CS amine group with a quaternary ammonium molecule, like GTMAC (Ahmed and Aljaeid, 2016; Liu et al., 2015).

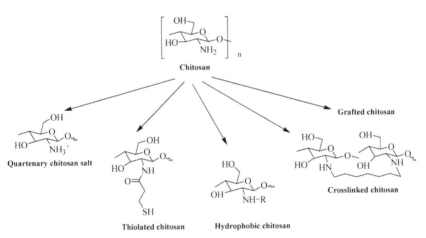

FIGURE 6.4 Broad classification of chitosan derivatives.

6.4.2 THIOLATED

Another commonly synthesized group of CS derivatives are the thiolated CSs. Thiolated CSs are synthesized when a modification is associated with the immobilization of thiol-bearing moiety onto the fundamental CS (Sakloetsakun and Bernkop-Schnürch, 2010). The three different conjugates under these types of derivatives are CS-thioglycolic acid conjugates (Constantia, 2001), chitosan-4-thio-butyl-amidine (chitosan-TBA) conjugates (Marta, Margit, and Paolo, 2004) and CS-cysteine conjugates (Zhang, Tang, and Yin, 2013). The synthesized derivatives indicated several properties such as enhanced mucoadhesive drug delivery and improved permeation effect. It has also been reported that due to the formation of inter and intramolecular disulfide bonds when in contact with physiological fluids, these types of derivatives are preferred for their *in situ* gelling characteristics. The reliable cohesiveness associated with thiolated CS makes it a favorable derivative for *in vitro* and *in vivo* investigations (Ahmed and Aljaeid, 2016).

6.4.3 HYDROPHOBIC

A new group of macromolecules that have gained industrial importance is the hydrophobic associating water-soluble polymers (Kumar et al., 2004). These types of chemically modified derivatives indicate improved stability due to the reduction in the matrix hydration process (Ahmed and Aljaeid, 2016). Synthesis of hydrophobic CSs occurs when there is a substitution with alkyl chains of adequate length (minimum six carbon atoms). However, certain parameters are responsible for the hydrophobicity of the derivative such as the nature of the hydrophobic sites, the number of hydrophobic sites, the concentration of the polymer, the temperature and the ionic strength of the medium (Desbrières, Martinez and Rinaudo, 1996). Hydrophobic CS can be broadly categorized into acylated and alkylated derivatives. N-Alkylated CS is prepared by the reaction of amino groups with aldehydes or ketones (the Schiff reaction) which leads to the formation of imines. The reaction further progresses by the reduction of the formed imines to amines with borohydride or sodium cyanoborohydride (Aranaz, Harris and Heras, 2010). Acylation has a broader range of possibilities due to the substitution of hydrophobic groups to the N-atom or the O-atom or to both simultaneously. This gives rise to N-acylated, O-acylated, and N,O-acylate CS. The N-acylated CS derivatives are synthesized by the interaction of the amine groups with anhydrides, acid chlorides, carboxylic acids, etc. (Philippova and Korchagina, 2012). Some of the common examples of hydrophobic CSs are CS succinate and CS phthalate. Other alternate ways to initiate hydrophobic groups to CS can be achieved by acylation using sodium salts of medium-chain fatty acids, fatty acid chlorides, and lactones, which lead to improving the permeability of the derivatives (Ahmed and Aljaeid, 2016; Sonia and Chandra, 2011).

6.4.4 CROSSLINKED

One of the most frequently described methods of synthesizing derivatives is CS modification through crosslinking. This simple procedure was observed during the grafting procedure, when certain parts of CS underwent immense crosslinking, which led to obtaining a modified CS. The modified CS so obtained demonstrates very different properties from native CS. It has also been observed that permanent covalent networks are formed due to polymer crosslinking, resulting in the improved mechanical strength of the polymers, which may be due to free diffusion of water/bioactive materials. The main applications of covalently crosslinked CS are in-site specific sustained drug delivery through diffusion, and as permanent networks used in tissue engineering (Harish Prashanth and Tharanathan, 2006). Depending upon the interaction between the crosslinker and CS, crosslinking can be broadly categorized into physical crosslinking and chemical crosslinking (Szymańska and Winnicka, 2015).

6.4.5 GRAFTED

Another common approach to derivatization of a moiety is through chemical grafting. Grafting allows us to create hybrid entities, which combine natural polysaccharides with synthetic polymers. Grafted CS can be synthesized by connecting single or multiple species of blocks to the main CS chain. Depending upon the length, number, and molecular structure of the side chains, the new derivative imparts characteristic physical and chemical properties (Ahmed and Aljaeid, 2016). The most common methods of grafting are through radiation or enzymatic approaches (Jayakumar et al., 2005). Other parameters that contribute to the type of derivative synthesized are based on aspects such as the monomer concentration, the initiator concentration, the reaction temperature and time, etc. Some of the common strategies are grafting the CS backbone with hydrophilic polymers such as PEG.

It has indicated better solubility and improved biocompatibility with the host immune system (Andrade et al., 2011). The grafted CS also indicated better thermostability (Ho et al., 2015). Grafted CS's were also synthesized with galactose groups linked to the CS amine group. Graft copolymerization with polyvinyl and polyacrylic moieties is an effective method for inducing modifications to the physicochemical characteristics of CS (Bhavsar et al., 2017).

6.5 APPLICATION IN PHARMACY AND MEDICINE

CS is adaptable to numerous applications in the pharmaceutical and medical industry, owing to its distinctive characteristics such as biocompatibility, biodegradability, nontoxicity, antibacterial, and antioxidant properties (Islam, Bhuiyan and Islam, 2017). However, the characterization and processing of chitins are very hard due to its insolubility in ordinary solvents, which limits its applications. Therefore, derivatization of CS has been the new interest of researchers so as to enhance its property to disintegrate at neutral and alkaline pH conditions while maintaining its cationic property at the same time (Alves and Mano, 2008). Out of all the methods to modify the chemical structure of CS, graft copolymerization is being used the most as it improves adsorption and chelation properties with bacteriostatic effect while retaining its mucoadhesivity and biodegradability (Alves and Mano, 2008). Biodegradation of CS relies on various factors, for example, molecular weight, pH, degree of acetylation along with its fabrication process (Zargar et al., 2015). CS is an appropriate hemostatic agent owing to the blood CS interaction that initiates with adsorption of plasma protein on the CS, which is followed by adhesion and activation of platelets that leads to thrombus development (Chou, Fu, Wu and Yeh, 2003; Yeh and Lin, 2008). Nevertheless, no thrombogenic activity was observed by both CS oligomers as well as water-soluble CS, whereas sulfated CS showed blood anticoagulation properties. Hence, modification of sulfates helped in obtaining CS derivatives with better hemocompatibility (Zargar et al., 2015). Due to all the excellent properties possessed by CS and its derivatives, it has diverse applications not only in medicine and pharmacy but also as a biomaterial for its use in the treatment of wounds as well as in orthopedic materials, as discussed in Table 6.1 (Mano et al., 2007; Singla and Chawla, 2001).

TABLE 6.1 Principle Applications of Chitosan and its Derivatives

Application	Example	Reference
Biomaterial	• Treatment of wounds, ulcers, and burns	Peña, Izquierdo-Barba, Martínez and Vallet-Regí, 2006; Singla and Chawla, 2001
	• Tissue regeneration and restoration	
Medicine and Pharmacy	• Surgical sutures and bandages	Peniche, Argüelles-Monal, Peniche and Acosta, 2003; Zargar et al., 2015
	• Microcapsules, microspheres, tablets, etc.	
	• Dentistry	
Tissue Engineering	• Cell growth and proliferation	Lee et al., 2000; Zargar et al., 2015
	• Tissue regeneration and bone repair	
	• Porous 3D scaffolds	
Cosmetics	• Skin and hair care products	Hirano, 1996
Agriculture	• Plant protection against diseases	Galed, Fernández-Valle, Martínez and Heras, 2004; Hirano, 1996
	• Seed and fruit coating fertilizer	
	• Enhance plant-microorganism interaction	
Food and Nutrition	• Emulsifying, thickening, and stabilizing agent	Dodane and Vilivalam, 1998; Hirano, 1996; Ravi, 2000
	• Food preservative	
	• Natural flavor extender	
	• Color stabilization	
Artificial Skin	• Biodegradable template for synthesizing neodermal tissue	Zargar et al., 2015

6.5.1 TREATMENT OF WOUNDS

CS and its derivatives prove to be significant biomaterials for wound healing purposes as it is not only anti-microbial but also shows special properties such as good biodegradability, biocompatibility as with low immunogenicity (Cheung et al., 2015; Roller, 2000). It offers a 3D matrix, which promotes the growth of tissue by stimulating cell proliferation and activating macrophage activity, which in turn improves granulation and the arrangement of the healing tissues. It steps up the healing process by improving vascularization and produces less scarring because it slowly degrades into N-acetyl-β-D-glucosamine, which promotes the proliferation of fibroblasts and stimulates the synthesis of hyaluronic acid in addition to regulating the deposition of collagen.

Jayakumar et al., studied chitin and CS as a promising biomaterial in the field of wound and burns treatment (Jayakumar et al., 2010). Due to good compatibility with the adipose-derived stem cells, appropriate tensile strength, and porous nature, the nano-fibrous adhesive material derived from CS is used in surgeries to improve wound healing. Moreover, it was observed that CS has miraculous effects for wound healing applications for patients undergoing sinusitis operation, skin grafting, as well as plastic surgeries (Cheung et al., 2015).

Researchers have developed and studied many composite scaffolds for the treatment of wounds, for example, α-chitin/nanosilver scaffolds prepared by Madhumathi et al., (2010) had antibacterial as well as blood coagulating property. Moreover, Sudheesh et al., (2010) assessed and characterized β-chitin/nanosilver scaffolds for cell adhesion property and discovered that the chitin scaffolds embedded with nanosilver were perfect for treating wounds and burns (Kumar et al., 2010). Therefore, there are a lot of formulations out on the market in numerous derivatives of CS in the form of hydrogels, films, composites, etc. For instance, CS-based Hem Con bandages are widely known to treat external hemorrhage in both military and civilian prehospital areas (Bennett et al., n.d.). Similarly, Celox gauze offers emergency bleeding control; CS is positively charged and therefore attracts red blood cells, which carry a negative charge and then develop a clot around the wound. As illustrated in the diagram below (Figure 6.5), CS becomes sticky and swells upon coming into contact with blood, forming a soft gel-like clot which seals the skin and controls the bleeding (Weng, 2008). Sparks and Murray fabricated a film from CS and gelatin complex to be used as a dressing. The British Textile Technology Group patented the procedure to make dressings from CS, which was obtained from a nonanimal source, i.e., the micro-fungi (Elieh-Ali-Komi and Hamblin, 2016). It has been established that growth factors can be stabilized by heparin. Therefore, heparin along with water-soluble CS is being examined as a wound-healing matrix (Kweon, Song and Park, 2003). There are numerous CS-based commodities on the market, for instance, Chito-Seal, Chitopack, TraumaStat, etc. that promote wound healing (Cheung et al., 2015).

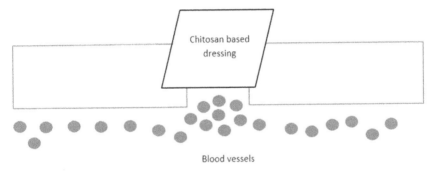

FIGURE 6.5 (See color insert.) Schematic representation of the working of the chitosan-based wound dressing.
Source: Cheung et al., (2015), https://creativecommons.org/licenses/by/4.0/

Other CS derivatives have application in wound healing purposes, for instance, the antibacterial activity of ethylene diamine tetraacetic acid grafted on CS, carboxymethyl chitosan (CHC-GG) with respect to only CS (Alves and Mano, 2008).

6.5.2 TISSUE ENGINEERING

The research in the area of tissue engineering primarily focuses on the employment of cells onto three-dimensional polymer matrices for restoring, regulating or enhancing biofunction by tissue (Dash et al., 2011). CS and anionic glycosaminoglycans (GAGs), which are found in the extracellular matrix of the skin, are found to have comparable physical and mechanical properties. The cationic nature of CS is useful for electrostatic interactions with the GAGs making it easier for cell attachment and regulating cell functions (Dash et al., 2011; Zargar et al., 2015). Therefore, CS and its derivatives seem promising candidates for application in tissue engineering because they not only but also degrade non-toxically without inflammation as the new tissue starts to form (Tuzlakoglu et al., 2004). CS with gelatin as well as collagen-based products has seen use in many regenerative applications such as bone tissue (Seol et al., 2004), cartilage, nerve (Cheng et al., 2003), liver (Kim et al., 2008), and trachea (Zargar et al., 2015).

CS with hydroxyapatite or CP glass has been used for clinical applications in bone tissue repair and regeneration (Di, Sittinger and Risbud, 2005). In addition, Zhang, and Zhang (2002) studied CS–calcium phosphate with CS sponge nested in calcium phosphate and observed that CS not only improved the mechanical strength by matrix reinforcement but also preserved the osteoblast phenotype. It has also been seen that porous CS mold permits osteoconduction and reduces any local inflammation (Levengood and Zhang, 2015). Hu et al., fabricated a CS hydroxyapatite composite with good bending modulus and mechanical strength and then proffered its application for fixating long bone fractures (Dash et al., 2011). The compressive characteristics have also been enhanced from 4 MPa to 11 MPa by developing and investigating macroporous CS-β tricalcium phosphate scaffolds for different pore sizes (Yin et al., 2003).

CS and chondroitin sulfate membrane have proved to aid chondrogenesis, making it an appropriate material for the repair of cartilage (Zargar et al., 2015). For developing articular cartilage, it is very important that the three-dimensional scaffolds have good degradation rate and significant porosity. Due to the similarity in structure between GAGs and CS, it was chosen as the scaffolding material for engineering the cartilage (Dash et al., 2011). Iwasaki et al., suggested the use of CS-alginate hybrid polymer for fabricating three-dimensional load-bearing scaffolds in cartilage regeneration as it showed increased tensile strength (Iwasaki et al., 2004). Lee et al., informed that CS-GAG scaffold loaded with transforming growth factor β1 (TGF- β1) not only had controlled release but also assisted regeneration of the cartilage tissue (Eun et al., 2004). The stability and mechanical properties of the collagen scaffold enhanced after adding CS (Taravel and Domard, 1996). Kim et al., (2003) showed chondrocyte proliferation through sustained release of TGF-β1 microspheres from a CS scaffold fabricated by the freeze-drying technique and hence, its utilization for treating cartilage conditions. Hoemann et al., exhibited that CS-glycerol phosphate implants mixed with blood enhance hyaline cartilage repair in ovine microfracture defects than in an ovine model alone (Kim et al., 2003). CS has also been used as a scaffolding material for hepatocytes due to its resemblance to the physical arrangement of GAGs present in the extracellular matrix of the liver (Li et al., 2003a, 2003b). Wang et al., (2003) developed a CS collagen matrix having good hepatocyte and blood compatibility using the crosslinking method. Nevertheless, it was necessary to develop a blood-contact device with anti-thrombogenic extracellular component as the formation of thrombus could reduce the efficiency of the membrane and hence, Wang et al., (Taylor et al., 2012) prepared a superior blood compatible matrix by gelating collagen and CS with heparin sodium containing ammonia, making it a prospective candidate for liver tissue engineering. Freier et al., (2005) prepared chitin hydrogel tubes which exhibited nerve cell adhesion and neurite outgrowth, showcasing the use of CS for preparation of scaffolds in the application of neural tissue engineering as well. Prabaharan et al., developed CS poly(L-lactic acid) scaffolds, which aimed its application as drug delivery system along with its use in tissue engineering. The drug release rate could be controlled by the amount of CS as well as the cross-link densities of the CS-PLLA hybrid (Alves and Mano, 2008). Additionally, CS-based grafted polymers such as poly(N-isopropylacrylamide) with chondrogenic mesenchymal stem cells have shown the likelihood in clinical applications such as cell therapy technologies especially in the treatment of vesicoureteric reflux (Gil and Hudson, 2004). Hence, CS and its derivatives offer significant scaffolds in many areas of tissue engineering, including its ease of fabrication and low cost (Yeh, Lin, and Lin, 2009).

6.5.3 PHARMACEUTICAL

CS and its derivatives have been utilized in numerous pharmaceutical applications as a drug carrier as well as a matrix in drug-releasing systems such as films, granules, and beads (Zargar et al., 2015). Many highly hydrophilic polymers such as gelatin, polyvinyl alcohol (PVA), etc. can be used to prepare membranes with different hydrophilic abilities as CS film exhibits low swelling in water (Ravi, 2000). Yao et al., (1995) prepared a crosslinked CS–gelatin polymer network and showed that the diffusion coefficient and relaxation time played a crucial role in pH-dependent drug release. Chandy et al., stated that the delivery of low molecular weight heparin through microspheres of CS-polyethylene glycol-alginate prevents thrombosis (Chandy et al., 2015). Chellat et al., (2000) studied the biodegradation behavior of CS-xanthan microspheres and reported it as a potential protective drug delivery system in the gastric and intestinal environment. By varying the crosslinking density of CS, the drug release can easily be changed in the transdermal drug delivery devices. The diffusion of macromolecules across the nasal barrier also makes CS nanoparticles as a suitable transporter for nasal drug doses (Zargar et al., 2015). CS with cationic charge is easily complexed with plasmid DNA as it carries a negative charge. The modified self-aggregates of CS work as a good drug delivery system–it creates charged complexes on mixing with plasmid DNA, which is beneficial for in vitro gene delivery into mammalian cells. The cationic character of CS has been essential in numerous useful purposes such as absorption improvement, bioadhesion, antimicrobial effect, etc., the cationic polymers of CS have application in the cosmetics industry. Chitopearl, another highly cationic derivative of CS developed by cross-linking porous beads of CS with a diisocyanate or diepoxy derivatives has application in chromatography and also as enzyme supports (Mourya and Inamdar, 2008). CS and its salts such as hydrochloride (HCl), glutamate show improved absorption for protein and peptide drugs. However, due to low solubility, CS, and its salts failed to improve permeability at pH 7.4, making it an unsuitable drug carrier for targeted drug delivery in certain areas of the intestine. TMC, the quaternized derivative of CS has indicated a much greater aqueous solubility demonstrating its application as an absorption enhancer for various test drugs (Mourya and Inamdar, 2008). TMC has also shown significant results in mucosal drug delivery (Alves and Mano, 2008). Modified CS centered materials have also been reported for intracellular gene therapy and DNA delivery being the most significant application of alkylated. Due to biocompatible characteristics of CS, Mao et al., (2004) showed that grafted CS materials could be utilized as a biomaterial in cardio-vascular engineering. Hence, various chemical modifications of CS have shown improved solubility in both neutral and alkaline solutions along with significant biomedical uses such as drug delivery, wound healing, as drug delivery, wound healing, and tissue engineering applications.

6.6 CONCLUSION

With the advancement in polymer engineering, a number of biopolymers are being discovered. Chitin and CS have established their significance by displaying their ability to be used in a wide array of applications. Due to its biocompatibility and biodegradability, CS has been an ideal candidate in the pharmaceutical domain. The further development in synthesizing the derivatives of CS has led to broadening the array of its pharmaceutical applications. This chapter gives a comprehensive overview of the general properties of CS, such as its structure and methods of extraction. It also discusses the broad classification of the derivatives synthesized from CS and finally elaborates on the recent applications of CS in wound healing, tissue engineering, and pharmaceutical domain.

KEYWORDS

- **biological activity**
- **chitosan**
- **chitosan derivatives**
- **pharmaceutical**
- **polysaccharide**

REFERENCES

Abdou, E. S., & Nagy, K. S. E. M., (2007). Extraction and characterization of chitin and chitosan from local sources. *Bioresource Technology, 99*(5), 1359–1367.

Agarwal, S., Leekha, A., Tyagi, A., Kumar, V., Moin, I., & Verma, A., (2015). Versatality of chitosan: A short review. *Journal of Pharma Research, 4*(3), 125–134.

Ahmed, T. A., & Aljaeid, B. M., (2016). Preparation, characterization, and potential application of chitosan, chitosan derivatives, and chitosan metal nanoparticles in pharmaceutical drug delivery. *Drug Design, Development and Therapy, 10*, 483–507. https://doi.org/10.2147/DDDT.S99651 (Accessed on 3 June 2019).

Alves, N. M., & Mano, J. F., (2008). Chitosan derivatives obtained by chemical modifications for biomedical and environmental applications. *International Journal of Biological Macromolecules, 43*(5), 401–414. https://doi.org/10.1016/j.ijbiomac.2008.09.007 (Accessed on 3 June 2019).

Andrade, F., Goycoolea, F., Chiappetta, D. A., Das Neves, J., Sosnik, A., & Sarmento, B., (2011). Chitosan-grafted copolymers and chitosan-ligand conjugates as matrices for pulmonary drug delivery. *International Journal of Carbohydrate Chemistry, 2011,* 1–14. https://doi.org/10.1155/2011/865704 (Accessed on 3 June 2019).

Aranaz, I., Harris, R., & Heras, A., (2010). Chitosan amphiphilic derivatives: Chemistry and applications. *Current Organic Chemistry, 14*(3), 308–330. https://doi.org/10.2174/138527210790231919 (Accessed on 3 June 2019).

Bennett, B. L., Littlejohn, L. F., Kheirabadi, B. S., Butler, F. K., Kotwal, R. S., Dubick, M. A., & Bailey, J. A., (n.d.). Management of External Hemorrhage in Tactical Combat Casualty Care: Chitosan-Based Hemostatic Gauze Dressings TCCC Guidelines– Change, 13–15.

Bhavsar, C., Momin, M., Gharat, S., & Omri, A., (2017). *Functionalized and Graft Copolymers of Chitosan and its Pharmaceutical Applications: Expert Opinion on Drug Delivery* (Vol. 14). Taylor & Francis. https://doi.org/10.1080/17425247.2017.1241230 (Accessed on 3 June 2019).

Cai, J., Yang, J., Du, Y., Fan, L., Qiu, Y., & Li, J. K. J. F., (2006). Enzymatic preparation of chitosan from the waste *Aspergillus niger* mycelium of citric acid production plant. *Carbohydrate Polymers, 64*(2), 151–157.

Chandy, T., Rao, G. H. R., Wilson, R. F., & Das, G. S., (2015). Delivery of LMW heparin via surface coated chitosan/peg-alginate microspheres prevents thrombosis delivery of LMW heparin via surface coated chitosan/peg-alginate microspheres prevents thrombosis, *7544,* 87–96. https://doi.org/10.1080/10426500290095584 (Accessed on 3 June 2019).

Chellat, F., Tabrizian, M., Dumitriu, S., Chornet, E., Rivard, C. H., & Yahia, L. H., (2000). Study of Biodegradation Behavior of Chitosan–Xanthan Microspheres in Simulated Physiological Media, 592–599.

Cheng, M., Deng, J., Yang, F., Gong, Y., Zhao, N., & Zhang, X., (2003). Study on physical properties and nerve cell affinity of composite films from chitosan and gelatin solutions, *24,* 2871–2880. https://doi.org/10.1016/S0142–9612(03)00117–0 (Accessed on 3 June 2019).

Cheung, R. C. F., Ng, T. B., Wong, J. H., & Chan, W. Y., (2015). Chitosan: An update on potential biomedical and pharmaceutical applications. *Marine Drugs* (Vol. 13). https://doi.org/10.3390/md13085156 (Accessed on 3 June 2019).

Chou, T. C., Fu, E., Wu, C. J., & Yeh, J. H., (2003). Chitosan enhances platelet adhesion and aggregation. *Biochemical and Biophysical Research Communications, 302*(3), 480–483. https://doi.org/10.1016/S0006–291X(03)00173–6 (Accessed on 3 June 2019).

Constantia, E., & Kast, A. B. S., (2001). Thiolated polymers–thiomers: Development and *in vitro* evaluation of chitosan–thioglycolic acid conjugates. *Biomaterials, 22*(17), 2345–2352.

Dash, M., Chiellini, F., Ottenbrite, R. M., & Chiellini, E., (2011). Progress in polymer science chitosan — a versatile semi-synthetic polymer in biomedical applications. *Progress in Polymer Science, 36*(8), 981–1014. https://doi.org/10.1016/j.progpolymsci.2011.02.001 (Accessed on 3 June 2019).

Desbrières, J., Martinez, C., & Rinaudo, M., (1996). Hydrophobic derivatives of chitosan: Characterization and rheological behavior. *International Journal of Biological Macromolecules, 19*(1), 21–28. https://doi.org/10.1016/0141–8130(96)01095–1 (Accessed on 3 June 2019).

Di, A., Sittinger, M., & Risbud, M. V., (2005). Chitosan: A versatile biopolymer for orthopedic tissue-engineering, *26*, 5983–5990. https://doi.org/10.1016/j.biomaterials.2005.03.016 (Accessed on 3 June 2019).

Dodane, V., & Vilivalam, V. D., (1998). Pharmaceutical applications of chitosan. *Pharmaceutical Science and Technology Today*, *1*(6), 246–253. https://doi.org/10.1016/S1461-5347(98)00059-5 (Accessed on 3 June 2019).

Douglas, D. B., Rejane, C. G., Sergio, P. C., & Filho, O. B. G. A., (2011). Quaternary salts of chitosan: History, antimicrobial features, and prospects. *International Journal of Carbohydrate Chemistry*, *2011*. https://doi.org/10.1155/2011/312539.

Elieh-Ali-Komi, D., & Hamblin, M. R., (2016). Chitin and chitosan: Production and application of versatile biomedical nanomaterials. *International Journal of Advanced Research*, *4*(3), 411–427.

Eun, J., Eun, K., Chan, I., Jeong, H., Lee, S., Cho, H., & Chul, M., (2004). *Effects of the Controlled-Released TGF- b 1 from Chitosan Microspheres on Chondrocytes Cultured in a Collagen/Chitosan/Glycosaminoglycan Scaffold*, *25*, 4163–4173. https://doi.org/10.1016/j.biomaterials.2003.10.057 (Accessed on 3 June 2019).

Freier, T., Montenegro, R., Shan, H., & Shoichet, M. S., (2005). *Chitin-Based Tubes for Tissue Engineering in the Nervous System*, *26*, 4624–4632. https://doi.org/10.1016/j.biomaterials.2004.11.040 (Accessed on 3 June 2019).

Galed, G., Fernández-Valle, M. E., Martínez, A., & Heras, A., (2004). Application of MRI to monitor the process of ripening and decay in citrus treated with chitosan solutions. *Magnetic Resonance Imaging*, *22*(1), 127–137. https://doi.org/10.1016/j.mri.2003.05.006 (Accessed on 3 June 2019).

Gil, E. S., & Hudson, S. M., (2004). Stimuli-responsive polymers and their bioconjugates. *Progress in Polymer Science (Oxford)*, *29*(12), 1173–1222. https://doi.org/10.1016/j.progpolymsci.2004.08.003 (Accessed on 3 June 2019).

Harish, P. K. V., & Tharanathan, R. N., (2006). Crosslinked chitosan—preparation and characterization. *Carbohydrate Research*, *341*(1), 169–173. https://doi.org/10.1016/J.CARRES.2005.10.016 (Accessed on 3 June 2019).

Hirano, S., (1996). Chitin biotechnology applications. *Biotechnology Annual Review*, *2*(C), 237–258. https://doi.org/10.1016/S1387-2656(08)70012-7 (Accessed on 3 June 2019).

Ho, T. H., Le, T. N. T., Nguyen, T. A., & Dang, M. C., (2015). Poly(ethylene glycol) grafted chitosan as new copolymer material for oral delivery of insulin. *Advances in Natural Sciences: Nanoscience and Nanotechnology*, *6*(3). https://doi.org/10.1088/2043-6262/6/3/035004 (Accessed on 3 June 2019).

Imen, H., & Fatih, O. J. R., (2016). Industrial applications of crustacean by-products (chitin, chitosan, and chitooligosaccharides): A review. *Trends in Food Science & Technology*, *48*, 40–50.

Islam, S., Bhuiyan, M. A. R., & Islam, M. N., (2017). Chitin and chitosan: Structure, properties and applications in biomedical engineering. *Journal of Polymers and the Environment*, *25*(3), 854–866. https://doi.org/10.1007/s10924-016-0865-5 (Accessed on 3 June 2019).

Iwasaki, N., Yamane, S., Majima, T., & Kasahara, Y., (2004). *Feasibility of Polysaccharide Hybrid Materials for Scaffolds in Cartilage Tissue Engineering : Evaluation of Chondrocyte Adhesion to Polyion Complex Fibers Prepared from Alginate and Chitosan*, 828–833.

Jayakumar, R., Menon, D., Manzoor, K., Nair, S. V., & Tamura, H., (2010). Biomedical applications of chitin and chitosan-based nanomaterials: A short review. *Carbohydrate*

Polymers, 82(2), 227–232. https://doi.org/10.1016/j.carbpol.2010.04.074 (Accessed on 3 June 2019).

Jayakumar, R., Prabaharan, M., Reis, R. L., & Mano, J. F., (2005). Graft copolymerized chitosan–Present status and applications. *Carbohydrate Polymers, 62*(2), 142–158. https://doi.org/10.1016/j.carbpol.2005.07.017 (Accessed on 3 June 2019).

Jo, G. H., Jung, W. J., Kuk, J. H., Oh, K. T., & Kim, Y. J. P. R., (2008). Screening of protease-producing Serratia marcescens FS-3 and its application to deproteinization of crab shell waste for chitin extraction. *Carbohydrate Polymers, 74*(3), 505–508.

Kato, Y., Onishi, H., & Machida, Y., (2003). Application of chitin and chitosan derivatives in the pharmaceutical field. *Current Pharmaceutical Biotechnology, 4*(5), 303–309. https://doi.org/10.2174/1389201033489748 (Accessed on 3 June 2019).

Kaur, S. D. G., (2014). The versatile biopolymer chitosan: potential sources, evaluation of extraction methods and applications. *Crit. Rev. Microbiol., 40*(2), 155–175.

Kean, T., Roth, S., & Thanou, M., (2005). Trimethylated chitosans as non-viral gene delivery vectors: Cytotoxicity and transfection efficiency. *J. Control Release, 103*(3), 643–653.

Khor, E., (2001). Chitin: Fulfilling a Biomaterials Promise. Elsevier Ltd., Amsterdam, 1st edition. pp. 136+xi, ISBN 0080440185, https://doi.org/10.1016/S0142-9612(02)00112-6.

Kim, I., Seo, S., Moon, H., Yoo, M., Park, I., Kim, B., & Cho, C., (2008). Chitosan and its derivatives for tissue engineering applications, *26*, 1–21. https://doi.org/10.1016/j.biotechadv.2007.07.009 (Accessed on 3 June 2019).

Kim, S. E., Park, J. H., Cho, Y. W., Chung, H., Jeong, S. Y., Lee, E. B., & Kwon, I. C., (2003). *Porous Chitosan Scaffold Containing Microspheres Loaded with Transforming Growth Factor- b 1 : Implications for Cartilage Tissue Engineering, 91*, 365–374.

Kumar, D. P., Dutta, J., & Tripathi, V. S., (2004). Chitin and chitosan: Chemistry, properties and applications. *Journal of Scientific & Industrial Research, 63*, 20–31. https://doi.org/10.1002/chin.200727270 (Accessed on 3 June 2019).

Kumar, P. T. S., Abhilash, S., Manzoor, K., Nair, S. V., Tamura, H., & Jayakumar, R., (2010). Preparation and characterization of novel b -chitin/nanosilver composite scaffolds for wound dressing applications. *Carbohydrate Polymers, 80*(3), 761–767. https://doi.org/10.1016/j.carbpol.2009.12.024 (Accessed on 3 June 2019).

Kumar, R. M. N. V., Muzzarelli, R. A. A., Sashiwa, H., & Domb, A. J., (2004). Chitosan chemistry and pharmaceutical perspective. *Chemical Reviews, 104*, 6017–6084. https://doi.org/10.1021/cr030441b (Accessed on 3 June 2019).

Kweon, D., Song, S., & Park, Y., (2003). Preparation of water-soluble chitosan/heparin complex and its application as wound healing accelerator, *24*, 1595–1601. https://doi.org/10.1016/S0142-9612(02)00566-5 (Accessed on 3 June 2019).

Lee, Y. M., Park, Y. J., Lee, S. J., Ku, Y., Han, S. B., Klokkevold, P. R., & Chung, C. P., (2000). The bone regenerative effect of platelet-derived growth factor-BB delivered with a chitosan/tricalcium phosphate sponge carrier. *Journal of Periodontology, 71*(3), 418–424. https://doi.org/10.1902/jop.2000.71.3.418 (Accessed on 3 June 2019).

Levengood, S. L., & Zhang, M., (2015). Chitosan-based scaffolds for bone tissue engineering. *J. Mater Chem. B Mater. Biol. Med., 2*(21), 3161–3184. Chitosan-based. https://doi.org/10.1039/C4TB00027G.

Li, J., Pan, J., Zhang, L., & Yu, Y., (2003a). Culture of hepatocytes on fructose-modified chitosan scaffolds, *Biomaterials, 24*, 2317–2322. https://doi.org/10.1016/S0142–9612(03)00048–6 (Accessed on 3 June 2019).

Li, J., Pan, J., Zhang, L., Guo, X., & Yu, Y., (2003b). Culture of primary rat hepatocytes within porous chitosan scaffolds. *J Biomed Mater Res A. 67*(3), 938–943. https://doi.org/10.1002/jbm.a.10076"10.1002/jbm.a.10076.

Liu, P., Meng, W., Wang, S., & Sun, Y. A. M., (2015). Quaternary ammonium salt of chitosan: Preparation and antimicrobial property for paper. *Open Medicine, 10*(1), 473–478. https://doi.org/10.1515/med-2015–0081 (Accessed on 3 June 2019).

Madhumathi, K., Kumar, P. T. S., Abhilash, S., Sreeja, V., Tamura, H., Manzoor, K., & Jayakumar, R., (2010). *Development of Novel Chitin/Nanosilver Composite Scaffolds for Wound Dressing Applications*, 807–813. https://doi.org/10.1007/s10856–009–3877-z (Accessed on 3 June 2019).

Mano, J. F., Silva, G. A., Azevedo, H. S., Malafaya, P. B., Sousa, R. A., Silva, S. S., & Reis, R. L., (2007). Natural origin biodegradable systems in tissue engineering and regenerative medicine: Present status and some moving trends. *Journal of the Royal Society Interface, 4*(17), 999–1030. https://doi.org/10.1098/rsif.2007.0220 (Accessed on 3 June 2019).

Mao, C., Zhao, W. B., Zhu, A. P., Shen, J., & Lin, S. C., (2004). A photochemical method for the surface modification of poly(vinyl chloride) with O-butyrylchitosan to improve blood compatibility. *Process Biochemistry, 39*(9), 1151–1157. https://doi.org/10.1016/S0032–9592(03)00225–5 (Accessed on 3 June 2019).

Marta, R., Margit, H., & Paolo, C. A. B. S., (2004). Mucoadhesive thiolated chitosans as platforms for oral controlled drug delivery: Synthesis and *in vitro* evaluation. *European Journal of Pharmaceutics and Biopharmaceutics, 57*(1), 115–121.

Mini, B., & Josef, W. C. G., (2011). Effect of deproteination and deacetylation conditions on viscosity of chitin and chitosan extracted from Crangon crangon shrimp waste. *Biochemical Engineering Journal, 56*(1), 51–62.

Mourya, V. K., & Inamdar, N. N., (2008). Chitosan-modifications and applications: Opportunities galore. *Reactive and Functional Polymers, 68*(6), 1013–1051. https://doi.org/10.1016/j.reactfunctpolym.2008.03.002 (Accessed on 3 June 2019).

Peña, J., Izquierdo-Barba, I., Martínez, A., & Vallet-Regí, M., (2006). New method to obtain chitosan/apatite materials at room temperature. *Solid State Sciences, 8*(5), 513–519. https://doi.org/10.1016/j.solidstatesciences.2005.11.003 (Accessed on 3 June 2019).

Peniche, C., Argüelles-Monal, W., Peniche, H., & Acosta, N., (2003). Chitosan: An attractive biocompatible polymer for microencapsulation. *Macromolecular Bioscience, 3*(10), 511–520. https://doi.org/10.1002/mabi.200300019 (Accessed on 3 June 2019).

Philippova, O. E., & Korchagina, E. V., (2012). Chitosan and its hydrophobic derivatives: Preparation and aggregation in dilute aqueous solutions. *Polymer Science Series A, 54*(7), 552–572. https://doi.org/10.1134/S0965545X12060107 (Accessed on 3 June 2019).

Prabaharan, M., (2008). Review paper: Chitosan derivatives as promising materials for controlled drug delivery. *Journal of Biomaterials Applications, 23*(1), 5–36. https://doi.org/10.1177/0885328208091562.

Raabe, D., Al-Sawalmih, A., & Yi, S. B. F. H., (2007). Preferred crystallographic texture of alpha-chitin as a microscopic and macroscopic design principle of the exoskeleton of the lobster Homarus americanus. *Acta Biomaterialia, 3*(6), 882–895.

Raphael, R., Heloise, R., Anne, D. R., Christine, N. D., & Jerome, V. P., (2011). Chitosan and chitosan derivatives in drug delivery and tissue engineering. *Adv. Polym. Sci., 244*, 19–44.

Ravi, K. M. N. V., (2000). A review of chitin and chitosan applications. *Reactive and Functional Polymers, 46*(1), 1–27. https://doi.org/10.1016/S1381–5148(00)00038–9 (Accessed on 3 June 2019).

Rinaudo, M., (2006). Chitin and chitosan: Properties and applications. *Progress in Polymer Science, 31*(7), 603–632. https://doi.org/10.1016/J. PROGPOLYMSCI.2006.06.001 (Accessed on 3 June 2019).

Roller, S., (2000). *Antimicrobial Actions of Degraded and Native Chitosan Against Spoilage Organisms in Laboratory Media and Foods, 66*(1), 80–86.

Sadeghi, A. M. M., Dorkoosh, F. A., Avadi, M. R., Saadat, P., Rafiee-Tehrani, M., & Junginger, H. E., (2008). Preparation, characterization and antibacterial activities of chitosan, N-trimethyl chitosan (TMC) and N-diethylmethyl chitosan (DEMC) nanoparticles loaded with insulin using both the ionotropic gelation and polyelectrolyte complexation methods. *International Journal of Pharmaceutics, 355*(1–2), 299–306. https://doi.org/10.1016/j.ijpharm.2007.11.052 (Accessed on 3 June 2019).

Saikia, C., & Gogoi, P., (2015). Chitosan: A promising biopolymer in drug delivery applications. *Journal of Molecular and Genetic Medicine, 4*. https://doi.org/10.4172/1747–0862. S4–006 (Accessed on 3 June 2019).

Sakloetsakun, D., & Bernkop-Schnürch, A., (2010). Thiolated chitosans. *Journal of Drug Delivery Science and Technology, 20*(1), 63–69. https://doi.org/10.1016/S1773–2247(10)50007–8 (Accessed on 3 June 2019).

Seol, Y., Lee, J., Park, Y., Lee, Y., Lee, S., Han, S., & Chung, C., (2004). *Chitosan Sponges as Tissue Engineering Scaffolds for Bone Formation*, 1037–1041.

Singla, A., & Chawla, M., (2001). Chitosan: Some pharmaceutical and biological aspects--an update. *The Journal of Pharmacy and Pharmacology, 53*(8), 1047–1067. https://doi.org/10.1211/0022357011776441 (Accessed on 3 June 2019).

Sonia, T. A., & Chandra, P. S., (2011). Chitosan and its derivatives for drug delivery perspective. *Adv. Polym. Sci., 243*, 23–54. https://doi.org/10.1007/12_2011_117 (Accessed on 3 June 2019).

Spinelli, V. A., Laranjeira, M. C. M., & Fávere, V. T., (2004). Preparation and characterization of quaternary chitosan salt: Adsorption equilibrium of chromium (VI) ion. *Reactive and Functional Polymers, 61*(3), 347–352. https://doi.org/10.1016/j.reactfunctpolym.2004.06.010 (Accessed on 3 June 2019).

Szymańska, E., & Winnicka, K., (2015). Stability of chitosan–A challenge for pharmaceutical and biomedical applications. *Marine Drugs, 13*(4), 1819–1846. https://doi.org/10.3390/md13041819 (Accessed on 3 June 2019).

Taravel, M. N., & Domard, A., (1996). Collagen and its interactions with chitosan: III. Some biological and mechanical properties. *Biomaterials, 17*(4), 451–455. https://doi.org/10.1016/0142–9612(96)89663–3 (Accessed on 3 June 2019).

Taylor, P., Wang, X., Yan, Y., Lin, F., & Xiong, Z., (2012). *Journal of Biomaterials Science, Preparation and Characterization of a Collagen/Chitosan/Heparin Matrix for an Implantable Bioartificial Liver*, 37–41. https://doi.org/10.1163/1568562054798554 (Accessed on 3 June 2019).

Tuzlakoglu, K., Alves, C. M., Mano, J. F., & Reis, R. L., (2004). *Production and Characterization of Chitosan Fibers and 3-D Fiber Mesh Scaffolds for Tissue Engineering*

Applications, 811–819. https://doi.org/10.1002/mabi.200300100 (Accessed on 3 June 2019).

Wang, X. H., Li, D. P., Wang, W. J., Feng, Q. L., Cui, F. Z., Xu, Y. X., & Werf, M. V. D., (2003). Crosslinked collagen/chitosan matrix for artificial livers, *24*, 3213–3220. https://doi.org/10.1016/S0142–9612(03)00170–4 (Accessed on 3 June 2019).

Weng, M., (2008). The effect of protective treatment in reducing pressure ulcers for noninvasive ventilation patients. https://doi.org/10.1016/j.iccn.2007.11.005 (Accessed on 3 June 2019).

Yao, K. D., Yin, Y. J., Xu, M. X., & Wang, Y. F., (1995). Investigation of pH-sensitive drug delivery system of chitosan/gelatin hybrid polymer network. *Polymer International*, *38*(1), 77–82. https://doi.org/10.1002/pi.1995.210380110 (Accessed on 3 June 2019).

Yeh, C. H., Lin, P. W., & Lin, Y. C., (2009). Procedia chemistry chitosan microfiber fabrication using microfluidic chips of different sheath channel angles and its application on cell culture. *PROCHE*, *1*(1), 357–360. https://doi.org/10.1016/j.proche.2009.07.089 (Accessed on 3 June 2019).

Yeh, H. Y., & Lin, J. C., (2008). Surface characterization and *in vitro* platelet compatibility study of surface sulfonated chitosan membrane with amino group protection-deprotection strategy. *Journal of Biomaterials Science, Polymer Edition*, *19*(3), 291–310. https://doi.org/10.1163/156856208783720985 (Accessed on 3 June 2019).

Yin, Y., Ye, F., Cui, J., Zhang, F., Li, X., & Yao, K., (2003). Preparation and characterization of macroporous chitosan–gelatin-tricalcium phosphate composite scaffolds for bone tissue engineering. *J Biomed Mater Res A. 67*(3), 844–855. https://doi.org/10.1002/jbm.a.10153.

Zargar, V., Asghari, M., & Dashti, A., (2015). A review on chitin and chitosan polymers: Structure, chemistry, solubility, derivatives, and applications. *Chem. BioEng. Reviews*, *2*(3), 204–226. https://doi.org/10.1002/cben.201400025 (Accessed on 3 June 2019).

Zhang, J., Tang, C., & Yin, C., (2013). Galactosylated trimethyl chitosan–cysteine nanoparticles loaded with Map4k4 siRNA for targeting activated macrophages. *Biomaterials*, *34*(14), 3667–3677. https://doi.org/10.1016/J. Biomaterials.2013.01.079 (Accessed on 3 June 2019).

Zhang, Y., & Zhang, M., (2002). Calcium phosphate/chitosan composite scaffolds for controlled in vitro antibiotic drug release. *J Biomed Mater Res. 62*(3), 378–386. https://doi.org/10.1002/jbm.10312.

CHAPTER 7

Pharmaceutical Applications of Xanthan Gum

SUBHAM BANERJEE[1] and SANTANU KAITY[2]

[1]*Department of Pharmaceutics, National Institute of Pharmaceutical Education and Research (NIPER) – Guwahati, Assam, India*

[2]*Cadila Healthcare Limited, Ahmadabad, Gujarat, India*

ABSTRACT

Xanthan gum (XG) a high molecular weight, water-soluble, microbial exo-polysaccharide derived from *Xanthomonas campestris*. It has an extensive application in diverse fields and widely been utilized as an additive in various industrial and biomedical applications viz., agriculture, food, paints, textile, toiletries, petroleum recovery, materials construction, cosmeceuticals, etc. Moreover, it has also shown great potential in the pharmaceutical field, including drug delivery with certain modifications to achieve the desired application. Moreover, XG derived biomaterials are suitable for targeted drug delivery applications, where such modification will be well suited for pharmaceutical dosage form design. Therefore, in this chapter, the significant use of XG for various pharmaceutical applications, including its brief overall descriptions, chemical structure, development of interesting pharmaceutical formulations including drug delivery applications and possible modifications for various pharmaceutical applications are well highlighted herewith with future potentials.

7.1 INTRODUCTION

In this era of "go-green," naturally occurring materials are gaining more potential interest in various application-based fields to overcome the biological and environmental complexities over the synthetic counterparts. A considerable number of research and developmental activities are still going on throughout the globe to develop biopolymer-based materials for various applications like food, and its packaging, agriculture, textile, paints, toiletries, petroleum, materials of construction, cosmetics, etc. Tremendous activities are reported now and then in specified fields like drug delivery and biomaterials for various pharmaceutical applications. This chapter is also aimed with highlighting the potential application of Xanthan gum (XG), a naturally occurring biopolymer, in various fields of biomedical and pharmaceutical applications. XG and its potential applications were first discovered in the year of 1963 at National Center for Agricultural Utilization Research, which is a part of United States Department of Agriculture (USDA) (Margaritis and Zajic, 1978). XG was extensively come into focus due to its versatile nature in the early of 1964. XG is a natural microbial exo-polysaccharide with excellent rheological properties, which is obtained from the bacterium called *Xanthomonas campestris* (a gram-negative bacterium) found on cabbage plants. XG is a non-toxic, non-irritating, and non-sensitizing in nature, with an excellent biocompatible and biodegradable profile (Palaniraj and Jayaraman, 2011). It is already an approved biopolymer sanctioned by the United States Food and Drug Administration (USFDA) in the year of 1969 for the application in food additive without any specific quantity limitations (Kennedy and Bradshaw, 1984). This biopolymer is registered in the Code of Federal Regulations (CFR) for use as a stabilizer and as a thickener also. Thus, these inherent versatilities of this biopolymer have made itself a widely acceptable biomaterial of choice for food and various pharmaceutical applications.

As far as the history says and also we depicted earlier about this biopolymer was first discovered in cabbage plant and observed the bacterium called *Xanthomonas campestris* to produce an extracellular polysaccharide with an exciting physicochemical feature. Till then, a vast number of applications and modifications of this biopolymer for intended application has been studied, which made it the most lucrative commercial biopolymer for various pharmaceutical applications including biomedical and industrial field as well. It's diversified structural-functional group, and interesting physicochemical properties have made it a most acceptable biopolymer for advanced, modified, and targeted drug delivery systems also.

7.2   OVERVIEW OF XG

After dextran, XG belongs to the next class of microbial exo-polysaccharide, which was commercialized globally. As depicted earlier, XG is really a very safe, economic biocompatible and biodegradable natural polymer, which is devoid of any organ irritation. These inherent diversified properties of XG made it the most researched material for pharmaceutical and biomedical applications over synthetic polymers (Chang et al., 2009). Additionally, XG is also safe for oral administration for both as food additives or matrix system for drug carriers. It is previously reported that natural gums are well metabolized in the intestine by the enzymes available at intestinal microflora leading to degradation to their individual sugar components (Deshmukh and Aminabhavi, 2014).

However, depending on the source variability XG shows a few disadvantages like uncontrolled hydration, pH-dependent solubility, chances of bacterial contamination, altered rheological profile during long-term preservation, temperature sensitivity, etc. (Jani et al., 2009; Rana et al., 2011). Thus, a high number of attempts have been made to fix the problems associated with this pristine gum and to make it suitable and alternate biomaterials against the existing synthetic polymers. A brief description, features as well as chemical structure of Xanthan gum is given in Table 7.1.

7.3   XG IN PHARMACEUTICAL APPLICATIONS

Due to the various functional moieties in the XG chemical structure as depicted in Figure 7.1, with its exciting physicochemical features makes this gum for the designing, tailoring, and development of more acceptable pharmaceutical formulations. For instance, XG act as suspending and emulsifying agents for the preparation of stabilized suspensions and emulsions, respectively, and also in imparting viscosity in oral liquid dosage preparations. Moreover, XG acts beautifully as binding and disintegrating agent in several conventional dosage forms, gelling agent in semi-solid preparations, and may be used as an emollient and demulcent agent in cosmeceuticals.

Therefore, based on the pharmaceutical applications point of view, XG here categorized as (i) components for conventional dosage forms and (ii) biopolymer for advanced drug delivery applications. Our literature survey in Table 7.2 includes many relevant and exciting examples of XG (either

TABLE 7.1   Salient Features of Xanthan Gum

Description	Features
Nonproprietary Names	British Pharmacopoeia–Xanthan gum
	European Pharmacopoeia–Xanthani gummi
Synonyms	Corn sugar gum, Keltrol, Polysaccharide B-1459, Rhodigel, Xantural, *Gummi xanthanum*
IUPAC name	9H-Xanthene
CAS number	11138-66-2
Nature	A natural hydrophilic biopolymer, which is hygroscopic as well as a hydrocolloid in nature
Source	Commercially by aerobic submerged fermentation from bacterium exo-polysaccharides
Botanical name	*Xanthomonas campestris*
Family	Xanthomonadaceae
Molecular formula	$(C_{35}H_{49}O_{29})_n$
Molecular weight	High molecular weight exo-polysaccharides with branched polymeric chains approximately 2–50 million-g/mol or ranges from 2×10^6 to 20×10^6 Da.
Chemical structure	(*see* Figure 7.1)
Chemistry	1,4-linked β-D-glucose residues, having a tri-saccharide side chain attached to other D-glucosyl residues. Their backbone is similar to cellulose. The side chains are β-D-mannose-1,4-β-D-glucuronic acid-1,2- β-D mannose, where the internal mannose is mostly ortho-acetylated, and a 4,6-linked pyruvic acid ketal may substitute the terminal mannose
Appearance	*Physical State:* Solid, amorphous powder
	Color: White to cream or tan
	Odor: Feeble no desagreable
	Taste: Characteristics

TABLE 7.1 *(Continued)*

Description	Features
Solubility	Dissolves easily on stirring into the hot or cold water. XG is insoluble in most organic solvents.
Charge	Non-ionic natural gum
Shape	Branched polymeric chain
Specific gravity	1.600 at 25°C
Loss on drying	15.0%
pH	6.0–8.0 for a 1% w/v aqueous solution
Heat of combustion	14.6 J/g or 3.5 Cal/g
Melting point	270°C
Freezing point	0°C for a 1% w/v aqueous solution
Refractive index	1.333 for a 1% w/v aqueous solution
Calorific value	0.6 Kcal/g
Viscosity (dynamic)	1200–1600 mPas (1200–1600 cP) for a 1% w/v aqueous solution at 25°C
Stability	Aqueous solutions are mostly stable over a wide pH and temperature range. The chemical stability of XG is stable under normal conditions. The thermal stability is almost superior to most other water-soluble polysaccharides
Storage	Store in a cool and dry place. Keep container tightly closed
Biocompatibility and biodegradability	Excellent biocompatibility, intrinsic immunogenic ability and fully biodegradable
Regulatory status	USFDA inactive pharmaceutical ingredients and excipients in non-parenteral medicines

Deshmukh and Aminabhavi (2014); Petri (2015); Katzbauer (1998).

in alone or in combination with other biocompatible polymers) applications in pharmaceutical research and product development process.

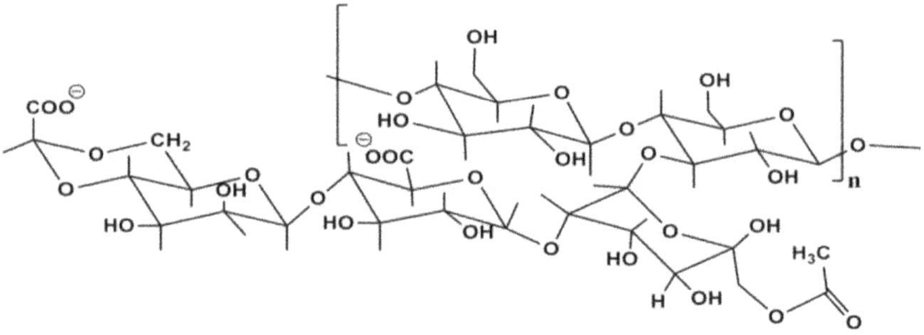

FIGURE 7.1   Chemical structure of XG.

7.3.1 COMPONENTS FOR CONVENTIONAL DOSAGE FORMS

Pharmaceutical components, excipients or ingredients for the development of conventional dosage forms are intended for the use of either as filler/ bulk or to improve the physicochemical properties of the final prepared dosage form by improving the stability profile with active pharmaceutical ingredients (APIs). Interestingly, XG attains those characteristics and thus gained particular attention as favorable components in making several conventional pharmaceutical formulations.

7.3.1.1 BINDING AND DISINTEGRATING AGENT

XG can act as a binding agent or binders due to its adhesive features in the presence of water. Therefore, XG is used as binder for binding the powder and to convert them into a cohesive mass of granules for tablet preparation, further it is then compressed and designed to form a solid dosage form like tablets, mucoadhesive buccal tablets, coated tablets, etc. (Ganesh et al., 2011; Talukdar and Kinget, 1995; Sinha et al., 2007). Examples of such preparations are given in Table 7.2 with the use of XG.

Moreover, XG can imbibe a considerable quantity of water and shows promising swelling potentiality in the presence of aqueous solution before it gets solubilize. Such exciting features convert XG to become

TABLE 7.2 Some Interesting Examples of XG in Pharmaceutical Applications Either Alone or in Combination with Other Biocompatible Polymers(s) for Potential Drug Delivery Applications

Polymer(s)	Drug(s)	Formulations	References
XG	Labetalol hydrochloride (HCl)	Mucoadhesive buccal tablets	Ganesh et al., 2011
	Caffeine, indomethacin, and sodium indomethacin.	Matrix tablets	Talukdar and Kinget, 1995
	Pentoxifylline	Tablets or gel	Mikac et al., 2010
	Chlorpheniramine maleate and theophylline	Matrix tablets	Dhopeshwarkar and Zatz, 1993
	Nicotine	Mucoadhesive patches	Abu-Huwaij et al., 2011
	Aspirin	Microspheres	Banerjee et al., 2013
XG and Boswellia gum (3:1)	5-FU	Compressed coated tablets	Sinha et al., 2007
XG and Guar gum (GG)	Dipyridamole	Floating matrix tablets	Patel and Patel, 2007
XG and GG (1:2)	5-FU	Matrix tablets	Sinha et al., 2004
XG and Galactomannan	Theophylline	Tablets	Vendruscolo et al., 2005
	Curcumin	Hydrogel	Koop et al., 2012
XG and Konjac glucomannan	Diltiazem hydrochloride	Tablets	Alvarez et al., 2008
	Cimetidine	Tablets or gel	Fan et al., 2008
XG and Chitosan (CS)	Metronidazole	Tablets	Eftaiha et al., 2010
XG, GG, and Hydroxypropyl methylcellulose (HPMC)	Propranolol hydrochloride	Tablets	Mughal et al., 2011
Kondagogu gum, GG, and XG	Glicazide	Mucoadhesive microcapsules	Mankala et al., 2011
XG and Lignin	Vanillin	Hydrogel	Raschip et al., 2011
XG, GG, and Poloxamer	Atropine sulfate	Hydrogel	Bhowmik et al., 2013

TABLE 7.2 *(Continued)*

Polymer(s)	Drug(s)	Formulations	References
XG and Modified starch	Ibuprofen (sodium salt), caffeine, sodium salicylate, vitamin B_{12}, Fluorescein isothiocyanate-inulin, and Fluorescein isothiocyanate-dextrans.	Hydrogel	Shalviri et al., 2010
XG and NaCMC	Diclofenac sodium	Hydrogel beads	Bhattacharya et al., 2013b
XG and Poly(vinyl alcohol) [PVA]		Hydrogel microspheres	Ray et al., 2010
	Ciprofloxacin hydrochloride		Bhattacharya et al., 2013a
XG, PVA, Na-hyaluronate, Carrageenan, and HPMC	Tetracycline HCL	Extruded paste	Obaidat et al., 2010
XG+Ethyl Cellulose	Diclofenac sodium	Microsponges	Maiti et al., 2011
XG+CS	C-phycocyanin	Chitosomes	Manconi et al., 2010
	Rifampicin	Liposomes	Manca et al., 2012
XG+HPMC+PVA	Zolmitriptan	Mucoadhesive buccal patches	Shiledar et al., 2014

an interesting disintegrating agent (Sinha et al., 2004; Vendruscolo et al., 2005). When XG administered into the solid dosage formulations as a disintegrating agent, it absorbs the surrounding dissolution medium and substantially build-up an internal osmotic pressure up to certain extent so that the compressed rigid physical integrity of the solid-dosage form disintegrates followed by dissolution into very fine particles.

7.3.1.2 HYDROPHILIC GELLING AGENTS

Natural gums show fascinating gelling property with water or aqueous medium. XG chains are highly branched and having wide numbers of connecting points. This property of such polymer helps them to form three-dimensional networks, which can imbibe a significant amount of water in its network structure. XG can produce such kind of swelled network structure even in the change of pH and temperature of the gelling media. Sometimes, the presence of other natural polysaccharides like other galactomannan or carrageenan can also induce the gelling phenomenon with XG (Prajapati et al., 2013).

The broad numbers of multiple sugar units in gums can also help to form inter and intramolecular hydrogen bonding or Van Der Walls interaction with water molecules. The concentration of the gum in the solution also plays a significant role to form the three-dimensional network structures with the desired viscosity. In most of the cases, a high weight percent of gum is required to form the network structure. However, in the case of XG, a decidedly less amount of gum is sufficient to produce the desired gelling effect (Koop et al., 2012).

7.3.1.3 EMULSIFYING AND SUSPENDING AGENT

XG can be used as an emulsifying and suspending agent. It used to work by stabilizing the thermodynamic interaction inside the system by preventing the recrystallization process. XG used to impart structural heterogeneity in the medium which in turn prevents the breakdown of the two miscible systems and helps to form thermodynamically stable biphasic liquid pharmaceutical preparations via reducing the surface and interfacial tension between these biphasic liquid media.

In suspension, XG used to increase the viscosity of the medium, and it forms a hydration layer around the suspended particles. Due to the highly branched structure, they used to form hydrogen bonding with the suspended particles. It prevents the aggregation of the particle and prevents subsequent settlement. The increase in viscosity of the medium plays a significant role to stabilize a suspension in the presence of natural gums like XG (Prajapati et al., 2013).

7.3.1.4   COATING AGENT

XG sometimes used as coating materials also. It used to act in two ways: a) it may act by protecting the APIs by its layers of coating or can retard the gastrointestinal tract (GIT) drug release; b) another way is to use XG as a granular and tablet coating material in compression coating system (Mitul et al., 2012). It is generally applied in combination with other natural polymeric counterparts for colorectal delivery (Niranjan et al., 2013). The primary advantage of using those natural gums specially XG as a coating material (Joshi et al., 2010) is to prevent early APIs release in upper GIT and to biodegrade in the specific region of the colon. Therefore, XG exhibit both excellent biocompatibility and biodegradability in a colonic region where it is used as a coating material for protecting the APIs or for prolonging the drug release from the designed matrix and to get easily digested by the colonic microflora due to its repetitive sugar units.

7.3.1.5   STABILIZING AGENT IN NANOPHARMACEUTICALS

XG can be used as an excellent stabilizing agent for nanoparticles synthesis to make the system stabilize physicochemically enough. Generally, it used to act by dual mechanism: (a) it can be adsorbed onto the surface of the nanoparticles and helps to create static repulsion between the synthesized particles; (b) XG can also be able to prevent the process of nanoparticles agglomeration by preventing them to come in contact with each other via improving the viscosity of the prepared dispersions or suspensions (Milas et al., 1990; Tiraferri et al., 2008; Comba and Sethi, 2009; Xue and Sethi, 2012).

XG can also be explored as a matrix for metal nanoparticles (gold and silver nanoparticles) synthesis and their subsequent stabilization.

In the case of gold nanoparticles, XG acts as both reducing agent and stabilizing agents due to abundant availability of many functional groups in its structural moiety and by metal-biosorption properties. XG at a very low concentration cannot only act as a reducing agent but capping agent also (Pooja et al., 2014). Such nanoparticles were found to be biocompatible since they are synthesized from the natural template. This kind of nanoparticles also showed high percentages of drug encapsulation efficiency (DEE). Hence, it can be said that XG is a well-studied natural gum for other metal nanoparticles synthesis than the other usual materials.

7.3.1.6 REDUCING AGENT IN NANOPHARMACEUTICALS

It can also be explored as potential reducing components for desired metal nanoparticles synthesis. A classic example like doxorubicin hydrochloride (HCl)-loaded gold nanoparticles which are synthesized from XG template through non-covalent interaction with excellent stability, and improved cytotoxicity can be used in cancer diseases like tumors, benign carcinoma, soft-tissue sarcomas and hematological malignancies (Pooja et al., 2014).

7.3.2 BIOPOLYMER FOR ADVANCED DRUG DELIVERY APPLICATIONS

Nowadays, a high number of potential researches are going on to explore the possibility of using natural gums as biodegradable and biocompatible advanced drug delivery systems. Natural gums are suitable to alternate over the synthetic counterparts who are generally applied for a better understanding of improving formulation stability, safety, and finally efficacy with desired bio palatability features by adopting cutting-edge modern technologies over the ongoing conventional strategies and therapies. On the same mechanism, researchers were already witnessed that XG can not only improve and ensure both the safety and efficacy power of an advanced novel drug delivery systems but also to regain overall therapeutic compliance of the patient community. Sometimes XG in single or mingle with other polymers (Table 7.2) showed exciting result while applied for controlled release delivery devices (Gilbert et al., 2013; Vianna-Filho et al., 2013). XG possess an exciting property to protect the

APIs in various pH conditions of GIT, low-pH stability, improved ionic strength, and controlling the release rate to give an optimal drug release in the desired destination. Thus, XG captured the researcher's attention and feasibility in gaining technological pursuit for the design of advanced healthcare delivery devices and systems.

7.3.2.1   SUSTAINED RELEASE MATRIX TABLETS

Tablet formulations are the most accepted and convenient method for delivering the active to its desired destination. Among various types of tablet formulations, prolong/sustained release, or controlled release matrix tablets gained an immense interest for controlling drug release rate from its designed matrix, therefore reducing the dose of the APIs and consequent drugs related adverse events. The principal mechanism behind sustained or controlled drug release from the designed tablets follows either by matrix erosion or by diffusion-controlled or swelling-dependent controlled release behavior (Ritger and Peppas, 1997).

   XG as native or its various modified forms can be used for developing matrix based suitable delivery devices. Modification of XG is done to impart desired functionality to the native polymer chain either by site-specific modification or via modifying the entire polymer backbone. This can individually satisfy the goals of targeted drug delivery systems. An investigation conducted by Mundargi et al. (2007) developed xanthan grafted copolymer of acrylamide and observed that grafting ratio could alter the release rate from the designed matrix (Mundargi et al., 2007). This kind of delivery system acts by adjusting the swelling and erosion mechanism of the total matrix system. XG and its combination with other polymers have been extensively well studied in designing matrix formulations for targeted applications. Specifically, reports for colon target drug delivery have been there with XG, modified XG and blending of XG with other natural gums like guar gum (GG), carrageenan, etc. (Verhoeven et al., 2006; Shaikh et al., 2011; Jackson and Ofoefule, 2011). The factors, which are majorly responsible for controlling the release are mainly, the concentration of XG in tablet preparations, the presence of other natural or synthetic polymer and the presence of engineered side chains incorporated onto the native gum backbone. Since XG is biodegradable; its concentration will not have that much impact on the biological system. So, it is safe to use in higher concentration as well. It is readily undergoing degradation in the presence of enzymes present in the colon (Jackson and Ofoefule, 2011; Jain et al., 2008; Asghar et al., 2009).

7.3.2.2 MATRIX PATCHES

Natural polymers like XG has been widely studied for developing matrix type patches to deliver the drug through the dermal layer for prolonged release of the drug. The drug release mechanism here also followed by swelling and erosion controlled release mechanism. The patches are fabricated by casting the gum and APIs with other additives into a thin film of uniform thickness. The factors, which influence the release rate, are the concentration of the rate-controlling polymer and the dosage uniformity throughout the thin film matrix. XG alone or in combination with other rate-controlling polymer also showed very promising results in various developed thin films based drug delivery systems. One promising combined mucoadhesive formulation mediated by XG and cationic GG for domperidone delivery was earlier reported by Singh et al., (2010). The bioavailability of such domperidone-loaded formulation was observed to be superior to that of oral conventional tablet formulation.

In another report, Abu-Huwaij et al., (2011) showed bilayer mucoadhesive patches of nicotine formulated by using XG in combination with carbopol 934 using ethyl cellulose as a backing membrane. The designed, fabricated patch showed well-tolerated compatibility with the APIs and the gum components. The mucoadhesive property of this formulation was promising, and the rate of drug release from the matrix was observed to be continued up to 10 hrs. A detail of the use of XG in the controlled delivery of active and small molecules has been tabulated in Table 7.2.

7.3.2.3 MICROPARTICLES

Multi-unit dosage forms like microparticles or microspheres gained a potential interest for several decades in designing controlled release systems. The average size ranges of microparticles are between 1–1000 µm. The major advantages of using such microparticle-based drug delivery systems are, a) it can deliver the APIs inside the GIT in a very consistent fashion, b) can ensure uniform drug absorption, c) local side effects can be minimized to a large extent, d) biodegradable, bioerodible, and finally biocompatible, etc. For these reasons XG is of choice of such kind of drug delivery systems development as it is biodegradable, bioerodible, and biocompatible, water-swellable, and easily be tailored to get modified multiparticulate based delivery systems (Banerjee et al., 2013).

Fabrication of microparticles based drug delivery system using XG is still an extensive research area to explore many more things. One of our research study, Ray et al., (2010) explored that interpenetrating polymer network (IPN) of XG, and synthetic polymer like polyvinyl alcohol (PVA) mediated controlled drug delivery systems can be intestinal pH-triggered. The microparticles were derived from emulsion crosslinking method and reported that the model drug (diclofenac sodium) could easily be pH-triggered at an intestinal region.

XG facilitated with superabsorbent (bentonite) synthetic polymer, i.e., PVA can also be an alternate approach for controlled release of multiparticulate-based drug delivery system. Another report, Bhattacharya et al., (2012) showed such kind of well-controlled release property following non-Fickian drug release kinetics.

XG, in combination with other natural gum like locust bean gum, can be an alternate approach than that of using synthetic polymer in combination. Desmukh et al., (2009) developed such a multiparticulate delivery system by inotropic-gelation technique. Such matrix type multiparticulate beads showed a well-controlled drug release from the polymer matrix to prolong the time of period. They found that an increase of the percentage of the natural polymer in the matrix, a multifold increase in DEE was observed.

7.3.2.4   IN SITU GELS

A novel type of advanced drug delivery system, which has come into the sight in present days, is *in situ* gel formulations. The sol system is composed of polymeric materials which used to turn into a gel when coming in contact with an external or internal shock like change in temperature, hydrogen ion concentration, ionic interaction or presence of various biological enzymes. Such types of formulations turned into gel-like structure inside our body and used to help in prolonging the release of the drug. Similarly, XG can be explored with a better understanding for such kind of application since it is said earlier that XG is biocompatible, biodegradable, and having exciting inherent properties of changing its physicochemical properties in the presence of various external or internal biological stimuli (Ray et al., 2010; Raschip et al., 2011).

7.3.2.5 NANOPARTICLES

Extensive research in the field of nanoparticulate based target-specific drug delivery systems has raised the urge of newer polymers for developing safer formulations. Use of natural gums for nanoparticles synthesis is still an arena of further translational health research. However, the concept or the hypothesis can be very much promising since the natural polymer does not create any complication related to biodegradability or biocompatibility issues.

Few reports are there of using natural polymers like XG being used as a reducing agent for the synthesis of metal nanoparticles. As said earlier, Pooja et al., (2014) reported the method of synthesis of gold nanoparticles using XG as a reducing agent. They observed that the so formed nanoparticles were biodegradable and biocompatible since the route of synthesis is based on the natural gum. So formed synthesized nanoparticles showed well time-dependent stable characteristics. Xue and Sethi (2012) already reported the feasibility of XG and GG in combination for making stabilized iron nanoparticles.

7.3.2.6 HYDROGELS

Hydrogel-based drug delivery systems for controlled release drug delivery is very much effective and promising since they can imbibe a large amount of water without dissolving or disintegrate the three-dimensional hydrogels network structure. Hydrogels can swell extensively and can help in prolonging the drug release as well. However, the modified forms of gums are more effective than that of its native counterparts for improving the DEE and for extending the drug release kinetics (Nishimura et al., 1993).

A classic example of such use of the modified natural polymer is reported by Gils et al., (2009). They published a hydrogel-based drug delivery system, which is enough porous in nature and composed of 2-hydroxyethyl methacrylate and acrylic acid grafted XG (Iijima et al., 2007). XG, as a native form, is unable to form a three-dimensional network in the presence of water. However, the network is built when the freeze-thaw process treats it at a specified temperature. The percentage of DEE plays a vital role in the process of drug release from the network structure of the hydrogel (Shalviri et al., 2010).

A dermatological hydrogel of a combined system of XG and galactomannan for curcumin delivery was well reported for anti-inflammatory activity (Koop et al., 2012). Another exciting use of hydrogel-based drug delivery systems is tailored IPN hydrogel beads for potential drug delivery applications (Bhattacharya et al., 2012). An interesting report showed a well absorbing capacity of polypropylene diglycidyl ether mediated crosslinked XG and lignin (Raschip et al., 2013). Reports are also there of delivering aloe vera by using chitosan (CS) and XG combined system for topical disorders (Chen et al., 2012).

7.3.2.7   LIPOSOMES

XG in the presence of poly-cationic polymers like CS can undergo extensive molecular complexation. Sometimes this property can be used to enhance vesicle stability of liposomes and also to improve payloads of delivered cargos. A functional coating of such a complex material observed to be effective for successful drug delivery of APIs by liposome-based formulation (Manca et al., 2012). XG in the presence of other natural gums like GG also showed good stability when used for β-carotene delivery by liposomal formulation approach (Toniazzo et al., 2014).

7.3.2.8   NIOSOMES

XG can improve the spreadability property of niosomal formulations. The APIs embedded inside the XG in niosomal formulations used to help in the formation of gel-like structure. In a useful report, Shinde et al., (2014) showed that the chances of leakage of enzymes also reduced drastically when a combined system comprises of XG is in niosomal formulations for controlled delivery from a gel base containing enzyme based niosomes. Thus, XG can help to modify the physicochemical properties, drug release pattern, and stability of enzyme-based niosomal gel formulations.

7.4 MODIFIED XG IN PHARMACEUTICAL APPLICATIONS

It is a well-known fact that the native form of natural gums has several demerits like uncontrolled swelling, chances of microbial contamination, reduced viscosity on long-term storage and many more, etc. To obtain the desired functionality and to improve the inherent demerits of the native form, physicochemical modification is essential (Figure 7.2). As far as the drug delivery point of concern, an intended native part modification is indispensable to meet the fundamental purpose than using the pristine one (Rana et al., 2011). Various chemically modified polymers and their specific pharmaceutical applications have been tabulated in Table 7.3. Possible chemical modification of natural gum can be done in multiple ways, namely, grafting, carboxymethylation, crosslinking, physical blend with another polymer(s) for synergistic effect or functional group substitution. It depends upon the particular need of the delivery systems that what specific type of modification is required to meet the intended goal. Among various modifications, some used to impart hydrophilicity and some modification used to make the polymer more hydrophobic. However, hydrophobic modifications are generally used in the paint industry rather than using as promising drug delivery carriers. The widely used method of imparting hydrophilicity in native gum is either by grafting or by carboxymethylation. The extent of hydrophilic nature depends on the degree of substitution (DS). In XG, such modification proved to be fruitful for their application in the drug delivery field. The primary target for modification site is the free hydroxyl (-OH) group, which is abundant in native polymer backbone of XG. The free -OH group acts as the soft target because of the steric hindrance, among other groups (Jiangyang et al., 2008). Some other natural gums, which were already modified by carboxymethylation process, are GG and konjac glucomannan (Thimma and Tammishetti, 2001). Polyelectrolyte or metal-ion crosslinked microbeads are the most straightforward approaches for developing controlled release drug delivery systems from the carboxymethylated natural gums.

Various ways can encounter crosslinking of modified natural gums. Crosslinking generally helps to arrest a significant amount of aqueous media in the network structure of the polymer complex. Drug delivery systems made by crosslinking the modified polymer did not collapse easily and continued to drug release for prolong time. This is the ideal need for a controlled release or targeted drug delivery system.

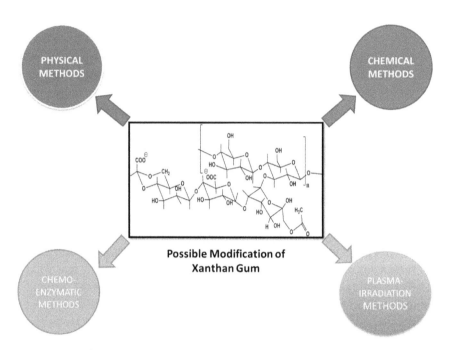

FIGURE 7.2   (See color insert.) Possible modification of XG chains used in pharmaceutical applications.

The choice of a specific crosslinker for designing a safe and effective controlled release drug delivery system is very much crucial. Among various crosslinkers, which are generally used for drug delivery applications, glutaraldehyde (GA) is extensively used and investigated. It is evident that an increase in GA concentration in a system will increase the crosslinking density of the polymer matrix and will subsequently reduce the inherent swelling ability of the component matrix structure. Such an exciting report has already been reported by Soppimath et al., (2000). They developed an IPN of GG and PVA using GA as a crosslinker and stated that the formation of acetal ring was responsible for the establishment of IPN and also for prolonged drug delivery application. In addition to this, the ionic crosslinked polymer network is also proved very effective. Di-valent or tri-valent metal ion crosslinked microbeads are abundant among the scientific reports intended for controlled release drug delivery.

Modification of natural gums by graft copolymerization method is very much effective for developing engineered biopolymers. Grafting of vinyl or acryl monomers onto the native polymer backbone used to increase the bulkiness of the total polymerized systems. These kinds of modified poly-mers used to possess high water retaining capacity without disruption of the network matrix structure. Graft copolymerization is generally executed by using various free radical initiators, which used to generate attachment site onto the native polymer backbone. Grafting of monomers can be done by using initiators like radiation, chemicals, or enzymes. Grafting of acryl-amide or polyacrylamide on the native polymer is extensively researched area in this context because this process is feasible and reproducible; large numbers of reports are there regarding controlled release drug delivery systems by using acrylamide grafted copolymers. Similarly, XG can also be modified according to its intended use employing grafting, carboxy-methylation or by simple crosslinked blend(s) with other natural gums to meet the specific goals of a controlled release drug delivery system (Badwaik et al., 2013). XG also exhibit a large number of hydroxyl (-OH) and carboxylic (-COOH) groups that can be tailored via different chemical means to obtain improved physicochemical properties of XG as biomate-rials (Bhattacharya et al., 2013b). Some common modification treatments are described in Table 7.3.

7.5 MISCELLANEOUS PHARMACEUTICAL APPLICATION

Thus XG can also be used in other alternate approaches with high confi-dence than that of the routine applications like controlled release drug delivery systems. XG can be used for stabilizing the suspension of barium sulfate, which is widely used as an X-ray contrast media. Sometimes it is used in cough syrup to prevent the settling of suspended materials by increasing the viscosity of the medium (Palaniraj et al., 2011).

Use of XG in cosmetics and toothpaste can prove the versatile nature of this polymer. Due to the inherent nature of its shear thinning property (Katzbauer, 1998), XG is widely used in toothpaste preparations. It helps in the easy extrusion of the paste from the dispensing tube. XG can also be used as a stabilizer and thickening agent in cosmetic products, which are cream or paste in nature. It is rare to XG as a therapeutic agent. No literature is found till date behind the support of this statement.

TABLE 7.3 Some Remarkable Examples of Modified XG in Pharmaceutical Applications

Modifications	Adopted strategies	Pharmaceutical applications	References
Physical method	Mechanical modification (high-pressure homogenization)	Structural and rheological properties changes	Eren et al., 2015
	Micro-fluidization (high-pressure treatment)	Controlled degradation without hampering its inherent chemical compositions	Laneuville et al., 2013
	Emulsification method	Lubrication behavior	Hamilton and Norton, 2016
	Blending with other polymers	XG modification with β-lactoglobulin via electrostatic attraction interaction showed inter-polymer complexes formation	Le and Turgeon, 2013
	Metal-ions complexation	Improved rheological properties and modified gelation behavior of XG	Nolte et al., 1992
Chemical method	Carboxymethylation	Enhanced DEE value and survival of chondrocyte cells for better therapeutic interventions for long-term cell therapies.	Mendes et al., 2012
		IPN beads of sodium carboxymethyl cellulose (SCMC) and sodium carboxymethyl XG by emulsion gelation technique using tri-valent ion crosslinker. *In-vitro* physicochemical drug release studies were found satisfactory.	Bhattacharya et al., 2013b
	Removal of existing functionality	Removal of acetyl groups from XG structure under higher alkaline conditions, the viscosity was observed to be increased.	Pinto et al., 2011
	Addition of new functionality	Exhibited improved salivary uptake for buccal drug delivery to treat sialorrhea.	Laffleur and Michalek, 2017
	Crosslinking	Developed pH-sensitive IPN microspheres by emulsion cross-linking method using GA to deliver an anti-inflammatory drug to the intestine.	Ray et al., 2010

TABLE 7.3 *(Continued)*

Modifications	Adopted strategies	Pharmaceutical applications	References
		Crosslinked the functional moieties via a covalent network for hydrogel synthesis and improved its mechanical stability.	Tao et al., 2016
		XG was derivatized to sodium carboxymethyl XG with aluminum ions to form microparticles from an aqueous template.	Maiti et al., 2008
	Grafting	Cyclodextrin acrylate crosslinked IPN hydrogels composed of XG maleate/N-isopropylacrylamide by a grafting-crosslinking method with a good potential for controlled release properties.	Hamcerencu et al., 2007
Chemo-enzymatic	Phosphorylase-copolymerization reactions	Amylose grafted-XG formed gel converted into a hydrogel with profound elastic property compared to native XG hydrogel.	Arimura et al., 2011
Plasma-irradiation	Cold-plasma technology	Cold-plasma treatment to graft primary amines (-NH2) followed by GA crosslinking may be used to obtain stable XG-based gels	Jampala et al., 2005

7.6   CONCLUSION AND FUTURE PERSPECTIVES

In conclusion, it can be said that XG is a polymer, which is versatile, can be used in several pharmaceutical applications in its native or its desired modified tailor-made form. The wide acceptability of this natural gum is that it is commercially available, nontoxic in nature, easily foldable to another form according to the desired intended use. These inherent properties have made it one of the most used and researched biopolymers among its class. As a drug delivery candidate, XG, and its modified forms can be entitled as lucrative excipients, which can be investigated with better understanding. In the field of drug delivery application, the potentiality of XG can be explored in the area of microparticulate drug delivery system, nanotechnology-based delivery systems and the area of biological products too.

Nevertheless, some other application fields related to XG remains still unexplored. Among them is XG based aerogels, nanocrystals because this gum is fascinating biomaterials with several interesting intrinsic physico-chemical properties, which can challenge and encourage researchers to create new architectures and functional systems based on it.

ACKNOWLEDGMENTS

The authors are thankful to their current institutions/organizations for writing this book chapter.

DISCLOSURES

No conflicts of interest.

KEYWORDS

- chemical modifications
- drug delivery
- pharmaceutical applications
- xanthan gum

REFERENCES

Abu-Huwaij, R., Obaidat, R. M., Sweidan, K., & Al-Hiari, Y., (2011). Formulation and *in vitro* evaluation of xanthan gum or carbopol 934-based mucoadhesive patches, loaded with nicotine. *AAPS Pharm. Sci. Tech., 12*, 21–27.

Alvarez-Mancenido, F., Landin, M., & Martinez-Pacheco, R., (2008). Konjac glucomannan/ xanthan gum enzyme sensitive binary mixtures for colonic drug delivery. *Eur. J. Pharm. Biopharm., 69*, 573–581.

Arimura, T., Omagari, Y., Yamamoto, K., & Kadokawa, J. I., (2011). Chemoenzymatic synthesis and hydrogelation of amylose-grafted xanthan gums. *Int. J. Biol. Macromol., 49*, 498–503.

Asghar, L. F., Chure, C. B., & Chandran, S., (2009). Colon-specific delivery of indomethacin: Effect of incorporating pH sensitive polymers in xanthan gum matrix bases, *AAPS Pharm. Sci. Tech., 10*, 418–429.

Badwaik, H. R., Giri, T. K., Nakhate, K. T., Kashyap, P., & Tripathi, D. K., (*2013*). Xanthan gum and its derivatives as a potential bio-polymeric carrier for drug delivery system. *Curr. Drug Deliv., 10, 587–600*.

Banerjee, S., Tiwari, A., Yadav, S. K., Bhattacharya, S. B., Dagaji, A. V., & Mondal, P., (2013). Influence on variation in process parameters for the design of xanthan gum facilitated ethyl cellulose microparticles for intestinal specific delivery. *Sci. Engg. Comp. Mat., 20*, 23–33.

Bhattacharya, S. S., Mazahir, F., Banerjee, S., Verma, A., & Ghosh, A., (2013a). Preparation and *in vitro* evaluation of xanthan gum facilitated superabsorbent polymeric microspheres, *Carbohydr. Polym., 98*, 64–72.

Bhattacharya, S. S., Mishra, A., Pal, D. K., Ghosh, A. K., Ghosh, A., Banerjee, S., & Sen, K. K., (2012). Synthesis and characterization of poly(acrylic acid)/poly(vinyl alcohol)-xanthan gum interpenetrating network (IPN) superabsorbent polymeric composites. *Polym. Plas. Tech. Engg., 51*, 876–882.

Bhattacharya, S. S., Shukla, S., Banerjee, S., Chowdhury, P., Chakraborty, P., & Ghosh, A., (2013b). Tailored interpenetrating network (IPN) hydrogel bead of sodium carboxymethyl cellulose and sodium carboxymethyl xanthan gum for controlled delivery of diclofenac sodium. *Polym. Plas. Tech. Engg., 52*, 795–805.

Bhowmik, M., Kumari, P., Sarkar, G., Bain, M. K., Bhowmick, B., Mollick, M. M. R., Mondal, D., Maity, D., Rana, D., Bhattacharjee, D., & Chattopadhyay, D., (2013). Effect of xanthan gum and guar gum on in situ gelling ophthalmic drug delivery system based on poloxamer-407. *Int. J. Biol. Macromol., 62*, 117–123.

Chang, P. R., Zhou, Z., Xu, P., Chen, Y., Zhou, S., & Huang, J., (2009). Thermoforming starch-graft polycaprolactone biocomposites via one-pot microwave assisted ring-opening polymerization. *J. Appl. Polym. Sci., 113*, 2973–2979.

Chen, S., & Zhang, L., (2012). *Skin Wound Repair Hydrogel and Preparation Method Thereof*, CN102698312-A.

Comba, S., & Sethi, R., (2009). Stabilization of highly concentrated suspensions of iron nanoparticles using shear-thinning gels of xanthan gum. *Water Res., 43*, 3717–3726.

Deshmukh, A. S., & Aminabhavi, T. M., (2014). Pharmaceutical applications of various natural gums. In: Ramawat, K., & Mérillon, J. M., (eds.), *Polysaccharides* (pp. 1–30). Springer.

Deshmukh, V. N., Jadhav, J. K., Masirkar, V. J., & Sakarkar, D. M., (2009). Formulation, optimization and evaluation of controlled release alginate microspheres using synergy gum blends. *Res. J. Pharm. Tec.*, *2*, 324–327.

Dhopeshwarkar, V., & Zatz, J. L., (1993). Evaluation of xanthan gum in the preparation of sustained release matrix tablets. *Drug Dev. Ind. Pharm.*, *19*, 999–1017.

Eftaiha, A. F., Qinna, N., Rashid, I. S., Al Remawi, M. M., Al Shami, M. R., Arafat, T. A., & Badwan, A. A., (2010). Bioadhesive controlled metronidazole release matrix based on chitosan and xanthan gum. *Mar. Drugs.*, *8*, 1716–1730.

Eren, N. M., Santos, P. H., & Campanella, O., (2015). Mechanically modified xanthan gum: Rheology and polydispersity aspects. *Carbohydrate Polymers, 134*, 475–484.

Fan, J. Y., Wang, K., Liu, M. M., & He, Z. M., (2008). *In vitro* evaluations of konjac glucomannan and xanthan gum mixture as the sustained release material of matrix tablet. *Carbohydr Polym.*, *73*, 241–247.

Ganesh, G. N. K., Manjusha, P., Gowthamarajan, K., Suresh, K. R., Senthil, V., & Jawahar, N., (2011). Design and development of buccal drug delivery system for labetalol using natural polymer. *Int. J. Pharm. Res. Dev.*, *3*, 37–49.

Gilbert, L., Loisel, V., Savary, G., Grisel, M., & Picard, C., (2013). *Stretching properties of xanthan, carob, modified guar and celluloses in cosmetic emulsions. Carbohydr. Polym., 93*, 644–650.

Gils, P. S., Ray, D., & Sahoo, P. K., (2009). Characteristics of xanthan gum-based biodegradable superporous hydrogel. *Int. J. Biol. Macromol.*, *45*, 364–371.

Hamcerencu, M., Desbrieres, J., Popa, M., Khoukh, A., & Riess, G., (2007). New unsaturated derivatives of Xanthan gum: Synthesis and characterization. *Polymer, 48*, 1921–1929.

Hamilton, I. E., & Norton, I. T., (2016). Modification to the lubrication properties of xanthan gum fluid gels as a result of sunflower oil and triglyceride stabilized water in oil emulsion addition. *Food Hydrocolloids, 55*, 220–227.

Iijima, M., Shinozaki, M., Hatakeyama, T., Takahashi, M., & Hatakeyama, H., (2007). AFM studies on gelation mechanism of xanthan gum hydrogels. *Carbohydr. Polym.*, *68*, 701–707.

Jackson, C., & Ofoefule, S., (2011). Use of xanthan gum and ethyl-cellulose in formulation of metronidazole for colon delivery. *J. Chem. Pharm. Res.*, *3*, 11–20.

Jain, S., Yadav, S. K., & Patil, U. K., (2008). Preparation and evaluation of sustained release matrix tablet of furosemide using natural polymers. *Res. J. Pharm. Tec.*, *1*, 374–376.

Jampala, S. N., Manolache, S., Gunasekaran, S., & Denes, F. S., (2005). Plasma-enhanced modification of xanthan gum and its effect on rheological properties. *J. Agri. Food Chem.*, *53*, 3618–3625.

Jani, G. K., Shah, D. P., Prajapati, V. D., & Jain, V. C., (2009). Gums and mucilages: Versatile excipients for pharmaceutical formulations. *Asian J. Pharm. Sci.*, *4*, 309–323.

Jiangyang, F., Wang, K., Liu, M., & He, Z., (2008). *In vitro* evaluations of konjac glucomannan and xanthan gum mixture as the sustained release material of matrix tablet. *Carbohydr. Polym., 73*, 241–247.

Joshi, M. G., Setty, C. M., Deshmukh, A. S., & Bhatt, Y. A., (2010). Gum ghatti: A new release modifier for zero order release in 3-layered tablets of diltiazem hydrochloride. *Indian J. Pharm. Educ. Res.*, *44*, 78–85.

Kar, R., Mohapatra, S., Bhanja, S., Das, D., & Barik, B., (2010). Formulation and *in vitro* characterization of xanthan gum-based sustained release matrix tables of isosorbide-5-mononitrate. *Iran. J. Pharm. Res.*, *9*, 13–19.

Katzbauer, B., (1998). Properties and applications of xanthan gum. *Polymer Degradation and Stability, 59,* 81–84.

Kennedy, J. F., & Bradshaw, I. J., (1984). Production, properties and applications of xanthan. *Progress in Industrial Microbiology, 19,* 319–371.

Koop, H. S., Da-lozzo, E. J., Freitas, R. A., Franco, C., Mitchell, D. A., & Silveira, J. L., (*2012*). Rheological characterization of a xanthan-galactomannan hydrogel loaded with lipophilic substances. *J. Pharm. Sci., 101,* 2457–2467.

Laffleur, F., & Michalek, M., (2017). Modified xanthan gum for buccal delivery-A promising approach in treating sialorrhea. *Int. J. Biol. Macromol., 102,* 1250–1256.

Laneuville, S. I., Turgeon, S. L., & Paquin, P., (2013). Changes in the physical properties of xanthan gum induced by a dynamic high-pressure treatment. *Carbohydr. Polym., 92,* 2327–2336.

Le, X. T., & Turgeon, S. L., (2013). Rheological and structural study of electrostatic cross-linked xanthan gum hydrogels induced by β-lactoglobulin. *Soft Matter, 9,* 3063–3073.

Maiti, S., Kaity, S., Ray, S., & Sa, B., (2011). Development and evaluation of xanthan gum-facilitated ethyl cellulose microsponges for controlled percutaneous delivery of diclofenac sodium. *Acta Pharm., 61,* 257–270.

Maiti, S., Ray, S., Mandal, B., Sarkar, S., & Sa, B., (2008). Carboxymethyl xanthan microparticles as a carrier for protein delivery, *J. Microencap., 24,* 743–756.

Manca, M. L., Manconi, M., Valenti, D., Lai, F., Loy, G., Matricardi, P., & Fadda, A. M., (2012). Liposomes coated with chitosan- xanthan gum (chitosomes) as potential carriers for pulmonary delivery of rifampicin. *J. Pharm. Sci., 101,* 566–575.

Manconi, M., Mura, S., Manca, M. L., Fadda, A. M., Dolz, M., Hernandez, M. J., Casanovas, A., & Díez-Sales, O., (2010). Chitosomes as drug delivery systems for C-phycocyanin: Preparation and characterization. *Int. J. Pharm., 392,* 92–100.

Mankala, S. K., Nagamalli, N. K., Raprla, R., & Kommula, R., (2011). Preparation and characterization of mucoadhesive microcapsules of gliclazide with natural gums. *S. J. Pharm. Sci., 4,* 38–48.

Margaritis, A., & Zajic, J. E., (1978). Biotechnology review: Mixing mass transfer and scale-up of polysaccharide fermentations. *Biotech. Bioengg., 20,* 939–1001.

Mendes, A. C., Baran, E. T., Pereira, R. C., Azevedo, H. S., & Reis, R. L., (2012). Encapsulation and survival of a chondrocyte cell line within xanthan gum derivative. *Macromol. Biosci., 12,* 350–359.

Mikac, U., Sepe, A., Kristl, J., & Baumgartner, S., (2010). A new approach combining different MRI methods to provide a detailed view on swelling dynamics of xanthan tablets influencing drug release at different pH and ionic strength. *J. Control. Rel., 145,* 247–256.

Milas, M., Rinaudo, M., Knipper, M., & Schuppiser, J. L., (1990). Flow and viscoelastic properties of xanthan gum solutions. *Macromolecules, 23,* 2506–2511.

Mitul, T. P., Jitendra, K. P., & Umesh, M. U., (2012). Assessment of various pharmaceutical excipients properties of natural *Moringa oleifera* gum. *Int. J. Pharm. Life Sci., 3,* 1833–1847.

Mughal, M. A., Iqbal, Z., & Neau, S. H., (2011). Guar gum, xanthan gum, and HPMC can define release mechanisms and sustained release of propranolol hydrochloride. *AAPS Pharm. Sci. Tech., 12,* 77–88.

Mundargi, R. C., Patil, S. A., & Aminabhavi, T. M., (2007). Evaluation of acrylamide-grafted-xanthan gum copolymer matrix tablets for oral controlled delivery of antihypertensive drugs. *Carbohydr. Polym., 69,* 130–141.

Niranjan, K., Shivapooja, A., Muthyala, J., & Pinakin, P., (2013). Effect of guar gum and xanthan gum compression coating on release studies of metronidazole in human fecal media for colon targeted drug delivery systems. *Asian J. Pharm. Clin. Res., 6*, 315–318.

Nishimura, S., Miyura, Y., Ren, L., Sato, M., Yamagishi, A., & Nishi, N., (1993). An efficient method for the synthesis of novel amphiphilic polysaccharide by regio- and thermo-selective modifications of chitosan. *Chem. Lett., 22*, 1623–1626.

Nolte, H., John, S., Smidsrød, O., & Stokke, B. T., (1992). Gelation of xanthan with trivalent metal ions. *Carbohyd. Polym., 18*(4), 243–251.

Obaidat, A. A., & Hammad, M. M., (2010). Sustained release of tetracycline from polymeric periodontal inserts prepared by extrusion. *J. Appl. Polym. Sci., 116*, 333–336.

Palaniraj, A., & Jayaraman, V., (2011). Production, recovery and applications of xanthan gum by *Xanthomonas campestris. J. Food Engg., 106*, 1–12.

Patel, V. F., & Patel, N. M., (2007). Statistical evaluation of the influence of xanthan gum and guar gum blends on dipyridamole release from floating matrix tablets. *Drug Dev. Ind. Pharm., 33*, 327–334.

Petri, D. F. S., (2015). *Xanthan gum: A versatile biopolymer for biomedical and technological applications. J. Appl. Polym. Sci., 132*, 420–435.

Pinto, E. P., Furlan, L., & Vendruscolo, C. T., (2011). Chemical deacetylation natural xanthan (Jungbunzlauer®). *Polímeros, 21*, 47–52.

Pooja, D., Panyaram, S., Kulhari, H., Rachamalla, S. S., & Sistla, R., (2014). Xanthan gum stabilized gold nanoparticles: Characterization, biocompatibility, stability and cytotoxicity. *Carbohyd. Polym., 110*, 1–9.

Prajapati, V. D., Jani, G. K., Moradiya, N. G., & Randeria, N. P., (2013). Pharmaceutical applications of various natural gums, mucilages and their modified forms, *Carbohyd. Polym., 92*, 1685–1699.

Rana, V., Rai, P., Tiwary, A. K., Singh, R. S., Kennedy, J. F., & Knill, C. J., (2011). Modified gums: Approaches and applications in drug delivery. *Carbohyd. Polym., 83*, 1031–1047.

Raschip, I. E., & Vasile, C., (2013). *Patent RO129066-A2.*

Raschip, I. E., Hitruc, E. G., Oprea, A. M., Popescu, M. C., & Vasile, C., (2011). *In vitro* evaluation of the mixed xanthan/lignin hydrogels as vanillin carriers. *J. Mol. Struct., 1003*, 67–74.

Ray, S., Banerjee, S., Maiti, S., Laha, B., Barik, S., Sa, B., & Bhattacharyya, U. K., (2010). Novel interpenetrating network microspheres of xanthan gum-poly(vinyl alcohol) for the delivery of diclofenac sodium to the intestine-*in vitro* and *in vivo* evaluation. *Drug Deliv., 17*, 508–519.

Ritger, P. L., & Peppas, N. A., (1997). A simple equation for description of solute release: II. Fickian and anomalous release from swellable devices. *J. Control Release, 5*, 37–42.

Shaikh, A., Shaikh, P., Pawar, Y., & Kumbhar, S., (2011). Effect of gums and excipients on drug release of ambroxol hydrochloride sustained release matrices. *J. Curr. Pharm. Res., 6*, 11–15.

Shalviri, A., Liu, Q., Abdekhodaie, M. J., & Wu, X. Y., (2010). Novel modified starch-xanthan gum hydrogels for controlled drug delivery: Synthesis and characterization. *Carbohydr. Polym., 79*, 898–907.

Shiledar, R. R., Tagalpallewar, A. A., & Kokare, C. R., (2014). Formulation and *in vitro* evaluation of xanthan gum-based bilayered mucoadhesive buccal patches of zolmitriptan. *Carbohydr. Polym., 101*, 1234–1242.

Shinde, U. A., & Kanojiya, S. S., (2014). Serratiopeptidase niosomal gel with potential in topical delivery. *J. Pharmaceutics.*, Article ID 382959.

Singh, M., Tiwary, A. K., & Kaur, G., (2010). Investigations on interpolymer complexes of cationic guar gum and xanthan gum for the formulation of bioadhesive films. *Res. Pharm. Sci.*, *5*, 79–87.

Sinha, V. R., Mittal, B. R., Bhutani, K. K., & Kumaria, R., (2004). Colonic drug delivery of 5-fluorouracil: An *in vitro* evaluation. *Int. J. Pharm.*, *269*, 101–108.

Sinha, V. R., Singh, A., Singh, S., & Binge, J. R., (2007). Compression coated systems for colonic delivery of 5-fluorouracil. *J. Pharm. Pharmacol.*, *59*, 359–365.

Soppimath, K. S., Kulkarni, A. R., & Aminabhavi, T. M., (2000). Controlled release of antihypertensive drug from the interpenetrating network poly(vinyl alcohol) guar gum hydrogel microspheres. *J. Biomater. Sci. Polym.*, *11*, 27–43.

Talukdar, M. M., & Kinget, R., (1995). Swelling and drug release behavior of xanthan gum matrix tablets. *Int. J. Pharm.*, *120*, 63–72.

Tao, Y., Zhang, R., Xu, W., Bai, Z., Zhou, Y., Zhao, S., Xu, Y., & Yu, D., (2016). Rheological behavior and microstructure of release-controlled hydrogels based on xanthan gum crosslinked with sodium trimetaphosphate. *Food Hydrocolloids*, *52*, 923–933.

Thimma, R., & Tammishetti, S., (2001). Barium chloride crosslinked carboxymethyl guar gum beads for gastrointestinal drug delivery. *J. Appl. Polym. Sci.*, *82*, 3084–3090.

Tiraferri, A., Chen, K. L., Sethi, R., & Elimelech, M., (2008). Reduced aggregation and sedimentation of zero-valent iron nanoparticles in the presence of guar gum. *J. Colloid Interface Sci.*, *324*, 71–79.

Toniazzo, T., Berbel, I. F., Cho, S., Favaro-Trindade, C. S., Moraes, I. C. F., & Pinho, S. C., (2014). β-carotene-loaded liposome dispersions stabilized with xanthan and guar gums: Physico-chemical stability and feasibility of application in yogurt. *LWT-Food Sci. Technol.*, *59*, 1265–1273.

Vendruscolo, C. W., Andreazza, I. F., Ganter, J. L. M. S., Ferrero, C., & Bresolin, T. M. B., (2005). Xanthan and galactomannan (from *M. scabrella*) matrix tablets for oral controlled delivery of theophylline. *Int. J. Pharm.*, *296*, 1–11.

Verhoeven, E., Vervaet, C., & Remon, J. P., (2006). Xanthan gum to tailor drug release of sustained-release ethylcellulose mini-matrices prepared via hot-melt extrusion: *In vitro* and *in vivo* evaluation, *Eur. J. Pharm. Biopharm.*, *63*, 320–330.

Vianna-Filho, R. P., Petkowicz, C. L., & Silveira, J. L., (2013). Rheological characterization of O/W emulsions incorporated with neutral and charged polysaccharides. *Carbohydr. Polym.*, *93*, 266–272.

Xue, D., & Sethi, R., (2012). Viscoelastic gels of guar and xanthan gum mixtures provide long-term stabilization of iron micro- and nanoparticles. *J. Nanoparticle Res.*, *14*, 1239.

Index

1

1,2,4-triazoles, 114
1-ethyl 3-(3-dimethylaminopropyl) carbodiimide (EDC), 9, 93, 104, 105, 128

2

2-acetamido-2-deoxy-bD-glucopyranose, 142
2-acrylamido-2-methylpropane sulfonic acid, 81
2-amino-2-deoxyb-D-glucopyranose, 143
2-chloro-1-methylpyridinium iodide, 4
2-hydroxyethyl methacrylate and acrylic acid, 179

3

3,6-anhydrogalactose, 73
3'-azido-3'-deoxythymidine (AZT), 9, 10, 113
3D matrix, 152
3-dimensional (3D) gelled-networking, 43

5

5-aminosalicylic acid, 131

8

8-formyl-7-hydroxy-4-methyl coumarin solution, 28

ε

ε-polylysine
  alginate nanocomposite particle, 17
  sodium alginate, 17

α

α-1,4-linked 3,6-anhydro-L-galactose, 78
α-L-guluronic acid, 40, 45, 59

β

β-(1–4)-linked D-glucosamine, 142
β-1,3-linked-D-galactose, 78
β-carotene, 124, 125, 180
β-chitin/nanosilver scaffolds, 152
β-cyclodextrin, 118
β-D-galactopyranosyl units, 79
β-D-glucose, 88
β-D-glucuronate, 88
β-D-mannuronic acid, 45, 59

κ

κ-carrageenan, 18, 22, 83, 116, 123, 125, 126, 128, 130, 131

λ

λ-carrageenan, 112

A

Acetamide group, 2, 144
Acetylsalicylate, 113
Acidic environment, 24, 123, 142
Acrylic acid, 81, 101
Acryloyl phenylalanine, 8
Active pharmaceutical ingredients (APIs), 112, 170, 174, 176, 177, 180
Acyl groups, 90
Acylation, 148
Adipocytic stem cells, 102, 105
Aerogel, 5, 13, 14, 31
Agar
  Agar, 71–74, 78–81, 84, 85, 88
    applications, 80
    solutions, 79
  alginate composite hydrogel, 17
  fabrication diagram, 77
  materials, 71, 84, 85
Agaropectin, 73, 77, 78
Agarose, 73, 78, 125

Alginate, 2–17, 22, 31, 38–59, 84, 88, 100, 119, 122, 131, 154, 156
  concentration, 41, 49
  hyaluronan
    composite, 5
    sponge, 6
  hydrogel, 3, 84
  matrix system, 48
  nanoparticles, 51
  poly acrylamide hydrogel, 8
  povidone iodine film, 15
  tetrabutylammonium hydroxide, 4
Alginates properties,
  bioadhesion, 45
  biocompatibility and immunogenicity, 44
  cross-linking and sol-gel transformation behavior, 42
  solubility, 41
  viscosity, 41
Alginic acid, 2, 38–40, 47, 48, 57, 59
Allopurinol mouthwash, 122
Alopecia treatment, 11
Aluminum alginate, 45
Ammonium
  glycyrrhetinic acid, 4
  persulphate, 8, 13
Amoxicillin system, 99
Amphiphilic test drugs, 125
Amphotericin B, 12, 13
Angiogenesis, 103, 113
Angiogenic compound, 113
Anionic groups, 91
Anti-adhesive properties, 104
Antibacterial, 24, 122, 150, 152, 153
Antibiotics, 71
Anticancer
  agent, 24
  drug delivery system, 27
Anticancer molecules, 27
Anticoagulant activity, 120
Anticoagulation action, 105
Antidiabetic activity, 11
Anti-HIV
  activity, 112, 113, 132, 133
  properties, 113
Anti-inflammatory
  agent, 18, 114

assays, 113
  drug, 115, 184
  effect, 104, 113
Antimicrobial
  activity, 10, 126
  films, 80
  test, 80
Antioxidant properties, 18, 38, 150
Antiseptic treatment, 15
Anti-thrombogenic extracellular component, 155
Antitubercular drugs, 51, 55
Antiviral activity, 112, 119
Aqueous
  medium, 41, 173
  solution, 11, 41, 51, 54, 169, 170
Ascophyllum nodosum, 2, 39
Ascorbic acid, 25, 26
Azotobacter species, 39

B

Bacillus subtilis, 24, 144
Bacterial
  contamination, 167
  sources, 39, 42
Bacterium, 80, 166, 168
Bamboo grills, 76
Beads/capsules, 98
Bentonite, 24, 25, 178
Beta
  cyclodextrin, 11
  hydroxy cyclodextrin, 10
Binding/disintegrating agent, 167, 170
Bioadhesion property, 45
Bioadhesive
  agents, 47
  characteristics, 45
  macromolecules, 119
Bioavailability, 31, 37, 45, 47, 51, 99, 100, 101, 116, 117, 119, 125, 177
Biocompatibility, 26, 38, 44, 46, 59, 71, 82–85, 101, 102, 104, 106, 125, 127, 141, 142, 145, 150, 152, 157, 169, 174, 179
Biodegradability, 2, 38, 71, 85, 104, 106, 141, 142, 145, 150, 152, 157, 169, 174, 179

Biodegradable, 31, 72, 84, 85, 112, 131, 166, 167, 169, 175–179
  films, 112
  natural polymer, 167
Biodegradation, 10, 82, 156
Biological
  activity, 141, 157
  substratum, 45
Biomedical
  adhesives, 141
  application, 88
  field applications, 102
    guided bone regeneration (GBR), 103
    surgery/wound healing, 104
    tissue engineering, 102
    treating intervertebral disorders, 103
Biomolecules, 92
Biomucoadhesive dosage forms, 38
Bionanocomposites, 112, 132, 133
  hydrogels, 126
Biopolymer, 106, 141, 142, 145, 166–168
  advanced drug delivery applications, 167, 175
Biosynthesis, 92, 93, 95
Biotechnology, 87, 132
Bismaeimide substitution, 8
Blood
  anticoagulation properties, 150
  clotting index, 6
  coagulating property, 152
  contact device, 155
  sugar level, 80
Bombyx mori, 7
Borohydride, 148
Bovine serum albumin (BSA), 3, 12, 17, 26, 58, 132
Bradykinin, 114
Breast cancer cell line (SK-BR-3), 27
Buccal
  films, 112, 120, 132
  membrane, 120

C

C. albicans, 80
C. codioides, 79
C. compressa, 114

C. coriacea, 79
C. crassa, 79
C. crinita, 114
C. irwinii, 79
C. obese, 79
C. racovitzae, 79
C. sedoides, 114
Calcitonin, 119
Calcium
  acetate, 48, 50
  alginate-hydroxypropylmethylcellulose, 5
  binding process, 43
  carbonate, 42, 51, 54, 58, 99, 144
    alginate nanoparticles, 51, 54
  cations, 44
  chloride solution, 10, 11, 13, 16, 22
  lactate, 144
  magnesium salts, 77
  salt, 2, 42, 77
Cancer chemotherapy, 122
Cancerous cells, 51
Carbohydrate, 73, 77
Carboxyl groups, 90
Carboxylate groups, 89, 90
Carboxylation, 93
Carboxylic acids, 148
Carboxymethyl
  cellulose, 20, 21, 47, 184
  groups, 93
Carboxymethylation, 93, 181, 183, 184
  thiolation, 93
Carboxymethylcellulose-(k) carrageenan composite, 21
Carrageenan, 1, 2, 17–24, 31, 83, 84, 88, 91, 111–120, 122, 123, 125–133, 172, 173, 176
Cation activity, 91
Cationic
  drug doxazosin, 130
  interaction, 91
  property, 150
Celecoxib, 115
Cell
  adhesion property, 152
  hyperplasia, 11
  migration, 102

mineralization, 129
therapy technologies, 155
transplantation, 46
Cellulose nanocrystals, 15
Celox gauze, 152
Centrifugation, 14, 24, 28, 76, 90, 94, 147
Chemical
bonding, 90
cross-linking procedures, 46
enzymatic processes, 141
modifications, 88, 93, 156, 186
scissoring (oxidation), 92, 93, 102
Chemotherapeutics, 71
Chitin, 2, 141–145, 152, 155
deacetylase, 145
Chitosan (CS), 1, 2, 11, 13, 23–29, 31, 52,
55, 58, 93, 99, 117, 118, 123, 126–129,
131, 141–150, 152–157, 171, 172, 180
bentonite nanofilm, 24
carrageenan complex, 23
derivatives, 145, 157
crosslinked, 149
grafted, 149
hydrophobic, 148
quaternary, 146
thiolated, 147
Hem Con bandages, 152
isolation, 143
chemical method, 144
enzymatic method, 144
structure, 142
thioglycolic acid, 147
Chlorhexidine, 14, 15
Chloroacetic acid, 93
Chloroform, 10, 41
Cholesterol absorption, 81
Chondrocyte proliferation, 93, 154
Chondrogenic
mesenchymal stem cells, 155
phenotype, 125
Chromatography, 156
Chronotherapeutic drugs, 117
Chymotrypsin, 12
Ciprofloxacin, 10, 172
Coating agent, 174
Code of Federal Regulations (CFR), 166
Colonic microflora, 174

Colon
specific drug releasing systems, 38
targeted delivery system, 21
Conformational transitions, 91
Conventional dosage forms, 167, 170
Cooling process, 42, 43
Cosmeceutical industry, 87
Cosmetics industry, 156
Cross-linking
process, 42, 43, 104
technique, 51
Crustaceans, 141, 142, 144
Curcumin, 1, 13, 14, 17, 55, 132, 180
delivery, 17
diglutaric acid, 13, 14
Curdiea angustata, 79
Cyclooxygenase-5/lipoxygenase, 5, 114
Cynodon daktylon, 113
Cytochrome C, 12
Cytokines, 114
Cytotoxic
drugs, 121
effect, 13
Cytotoxicity, 9, 15, 27, 116, 125, 175

D

D- and L-galactopyranosyl residues, 78
Decolorization, 144
Degradation behavior, 92
Degree of substitution (DS), 146, 181
Demineralization, 144
Deproteinization, 144
Deprotonated carboxylate group, 8
Diammonium dihydroxy platinum (IV)
dichloride, 4
Dichloromethane, 8
Diclofenac sodium, 26, 52, 53, 178
Discrete oligoethyleneglycolation
(dPEGylation), 115
Dissolved oxygen and oxygen transfer
capacity (DOT), 95
Divalent cations, 89, 90, 92
Di-valent/tri-valent metal ion, 182
Dodecyltrimethylammonium chloride, 131
Doxazosin mesylate selective membrane
electrode, 130

Drug
  candidates, 37, 47–49
  delivery, 26, 27, 29, 31, 37, 38, 44–46,
    49, 51–57, 59, 72, 82, 84, 100, 106,
    111–113, 117, 120, 121, 127, 131,
    132, 142, 145, 148, 149, 155, 156,
    165–167, 175–184, 186
    applications, 165, 180, 182
    system, 54, 59, 145, 156, 165, 175,
      176, 180, 182, 183, 186
  encapsulation
    efficiency (DEE), 175, 178–180, 184
    loading efficiency, 13
  releasing systems, 38, 57

E

E. singularis activity, 114
Electrospun fibrous mat, 10
Electrostatic interaction, 91, 116, 127, 130,
  131, 153
Emulsifying/suspending agent, 173
Emulsion techniques, 118
Encapsulation efficiency, 9, 11–13, 16, 17,
  22–24, 26–29
Endothelial cells, 128
Enzymatic
  approaches, 149
  modifications, 141
  treatment, 144
Enzymes, 46, 57, 83, 94, 144, 167, 177,
  178, 180, 183
Epithelial colorectal adenocarcinoma cell
  line, 13
Ethanolic solution, 17, 26, 29
Ethyl alcohol, 41, 90
Ethylene glycol, 41, 120
Eucheuma spinosum, 120
Eunicella singularis, 113
Exopolysaccharides, 93
Extracellular matrices, 39, 46

F

Fermentation, 87, 88, 92, 94, 144, 168
Ferric chloride, 24
Ferrous chloride, 24

Ferulic acid, 25, 26
Fibrinogen, 120
Fibroblasts, 103, 104, 152
Fibronectin, 103, 105
Fickian distribution, 3
Filtration, 74, 75, 90, 94
Food chemicals codex (FCC), 72
Franz-cell diffusion apparatus, 123
Freeze-drying process, 120
Freezing-thawing method, 76
Furfurylamine, 8

G

Galactose groups, 150
Gantrez polymer, 132
Gastrointestinal
  mode, 16
  tract (GIT), 54, 100, 119, 126, 174, 176,
    177
Gastroretentive
  drug-releasing efficacy, 57
  floating
    delivery formulations, 54
    tablets, 57
Gel
  filtration chromatography, 94
  formation, 42, 122
  forming properties, 41
  melting temperature, 79
Gelatin, 21, 47, 48, 50, 73, 82, 83, 98, 103,
  104, 122, 128, 130, 153–155
  capsules, 47, 48, 50, 54, 98
  encapsulate, 21
  hydroxyapatite, 104
  types, 82
    gelatin A, 82
    gelatin B, 82
Gelation, 41, 42, 44, 49, 71, 78, 79, 82, 87,
  89–94, 99, 101, 102, 106, 131, 184
Gelidiella, 74
Gelidium amansii, 74, 75
Gellan, 53, 87–95, 98–106, 122
  chains, 101, 104
  fermentation, 93
  gum, 53, 87–95, 100, 101, 104, 122
  polymer chains, 90

Gelling
  agent, 46, 72, 167
  process, 43
Gelosa, 73
Gelrite, 94
Gene
  delivery systems, 38, 59
  therapy, 156
Glacial acetic acid, 5, 24, 28, 42
Glucose unit, 88, 90
Glucuronic acid, 78, 90, 94, 96, 168
Glutamic acidalginate conjugate, 8
Glutaric acid and glutaraldehyde (GA), 83,
  123, 128, 182, 184, 185
Glyceryl
  methacrylate, 3
  monostearate, 12
  trimethyl ammonium chloride
    (GTMAC), 147
Glycosaminoglycans (GAGs), 153–155
Glycosphingolipids, 88
Glycyrrhetinic acid, 4, 55
Golgi apparatus chains, 78
Gracilaria, 74, 76, 79, 81
Granulation, 50, 152
Griffithsin, 21
Guar gum (GG), 40, 44, 59, 88, 93, 153,
  171, 176, 177, 179–182
Guided bone regeneration (GBR), 103
Guluronic acid, 2, 39, 41–44, 59

H

Helicobacter pylori, 99
Hematological malignancies, 175
Hemocompatibility, 24, 150
Hemorrhage, 123, 152
Hepatic accumulation, 132
Hepatocyte, 155
Herpes simplex virus, 22
Hetro-polysaccharide, 87
High acyl gellan gum (HAGG), 90
Higuchi and first-order model, 13
Histamine, 114
Homogeneous matrix, 43, 44
Homopolysaccharides, 39
Hormones, 46, 57, 116

Human
  breast cancer cell lines, 13
  epidermal growth factor receptor, 27
  hepatocellular carcinoma, 13
  immunodeficiency virus, 8, 21, 112
  papilloma virus, 21
Hyaluronic acid, 104, 105, 152
Hydraulic presses, 76
Hydrochloric acid, 91, 144
Hydrochloride (HCl), 9, 29, 47, 50, 53, 56,
  99, 100, 115, 118, 156, 171, 175
Hydrocolloid material, 94
Hydrogel, 3–5, 7, 8, 10, 16, 18, 22, 23, 26,
  31, 40, 46, 52, 53, 58, 71, 81–85, 92, 93,
  101, 103–106, 112, 118, 122–128, 132,
  133, 152, 155, 179, 180, 185
Hydrogen bonding, 78, 173, 174
Hydrophilic
  gelling agents, 173
  matrix systems, 49
  polymeric materials, 47
  polymers, 47, 50, 149, 155
Hydrophilicity, 181
Hydrophobic
  cooperative effects, 126
  electrostatic interactions, 130
  groups, 148
  sites, 148
Hydrophobicity, 98, 130, 148
Hydroxy group, 146
Hydroxyapatite, 14, 104, 129, 154
Hydroxyl
  group, 146
  propyl methylcellulose, 47
  radical scavenging, 18
Hydroxyproline concentration, 15
Hydroxypropyl methylcellulose (HPMC),
  5, 50, 54, 57, 125, 171, 172
Hygroscopicity, 21
Hyperglycemia, 12

I

Ibuprofen, 6, 7, 47, 50, 115, 130
Imatinib mesylate, 8
Imidazolyl-1,3,4-oxadiazoles, 114
Immobilization, 83, 147

Immune
reaction, 125
system, 150
Immunodeficiency, 8, 21, 112
Immunogenicity, 44, 45, 59, 152
Indomethacin, 14, 16, 114, 115, 120, 121, 171
Inflammatory activity, 111, 180
Inorganic
materials, 92
salts, 94
sulfate, 73
Inotropic-gelation technique, 178
In-situ
gelling systems, 99
gels, 178
Inter
penetrating polymer network (IPN), 99, 178, 180, 182, 184, 185
polyelectrolyte complexes, 118
Intestinal delivery, 16, 21, 31
Intravaginal drug administration, 112
Inulin (molecular weight marker), 119, 172
Ionic
crosslinked methacrylated gellan gum (iGG-MA), 92
cross-linking, 40, 44, 101
emulsion-gelation technique, 49
gelation
covalent cross-linking techniques, 49
technique, 51
Ionization, 90, 118
Ionotropic gelation, 13, 15, 98, 123, 125, 128
Iota carrageenan, 2, 18, 21
Isoelectric point, 17, 43

K

Kaku-Kanten, 75
Kanten, 73
Kappacarrageenan, 21
Kelcogel, 94
Ketoprofen, 5, 123, 124
Ketorolac, 115
Kinetics, 5, 14, 22, 26, 28, 48, 82, 92, 101, 118, 123, 126, 178, 179
Konjac gum, 88

Korsmeyer Peppas model, 48
kinetic model, 7, 8

L

L. plantarum cells, 21
Lactic acid, 144, 155
Laminaria
digitata, 2
hyperborea, 2, 39
japonica, 2
L-galactopyranosyl residues, 78
L-glycerate, 90
L-glyceryl substituents, 88
Light irradiation, 92
Liposomal formulation, 111, 115, 180
Liposome-based formulation, 180
Low acyl gellan gum (LAGG), 90, 92
L-rhamnose, 87, 88
Lyophilization, 120
Lyophilized formulation, 14, 21
Lysine, 101, 113
Lysozyme, 12

M

Macrocystis pyrifera, 2, 39
Macromolecular
drugs, 46
structures, 130
Macrophage activity, 152
Magnetic κ-carrageenan, 23
Mannuronic acid, 2, 39, 40, 59
Matrix
erosion, 49, 176
formers, 47
hydration process, 148
patches, 177
sustained release agent, 98
tablets, 48, 50, 171, 176
Metal-biosorption properties, 175
Methacrylated gellan gum, 92
Methacrylation, 92
Methyl
alcohol, 41
iodide, 146, 147
Methylation, 78
Metronidazole, 52, 57

Microbes, 2, 93, 112
Microbial
    fermentation, 88
    property, 80
Microcrystalline cellulose, 116, 117
    binding properties, 117
    pellets, 116
Microfibril, 142
Microparticles, 49, 52, 54, 56, 100, 118,
    130, 131, 177, 178, 185
Minoxidil, 10, 11
Monomer
    concentration, 149
    units, 40, 42
Monosaccharide, 91
Monovalent sodium cations, 43
Mucoadhesive
    buccal tablets, 170
    drug deliveries, 45
    formulation, 6, 177
    property, 18, 93, 100, 101, 130, 177
Mucoadhesiveness, 18, 99, 100, 106
Mucor rouxii, 145
Mycobacterium tuberculosis, 15
Myoglobin, 12

N

N-(2-hydroxyl) propyl-3-trimethyl ammo-
    nium chitosan chloride (HTCC), 147
N-acetyl, 142, 143
    D-glucosamine, 142
    β-D-glucosamine, 152
N-acylated, 148
    CS derivatives, 148
N-alkyl/quaternary ammonium, 146
N-alkylated CS derivatives, 146
N-trimethyl chitosan (TMC), 147
N-vinyl pyrrolidone, 3
Na-carboxymethyl cellulose, 119
Nanocarriers, 51, 132
Nanocomplex, 14, 15
Nanocomposite, 13, 14, 17, 24, 25, 71,
    126, 127, 129,
Nanocrystalline, 15
    hybrid, 15
Nanodelivery, 132

Nanoformulation, 9, 11, 12, 28, 31, 131
Nanogels, 131
Nanohydrogels, 101
Nanoparticle, 4, 5, 8, 9, 11, 13, 17, 26–29,
    31, 51, 54, 55, 58, 59, 80, 101, 105, 111,
    113, 115, 125–127, 131, 132, 156, 174,
    175, 179
Nanopharmaceuticals, 175
Nasal
    administration, 119
    drug delivery, 112, 119
    formulations, 100
Nasolacrimal drainage, 99
Natural
    polymer-based beads, 98
    polymers, 31, 71, 72, 88, 106, 142, 177,
        179
    polysaccharides, 38, 57, 149, 173
Neomycin agar, 74
Neural tissue engineering, 155
Neuropeptides, 115
Neutral polymer, 77
Neutrophil, 115
N-hydroxysuccinimide, 9
Niosomal formulations, 180
Niosomes, 180
N-isopropyl acrylamide, 8, 101
    N,N-bismethylenebisacrylamide, 8
Nitric acid, 144
Nitrosalicylaldehyde, 26
Non-Fickian mechanism, 100
Nonenzymatic hydrolytic degradation, 24
Nucleus pulposus, 92, 103

O

O-acetate bound, 90
O-acetyl moieties, 94, 96
Ophthalmic formulations, 99
Oral
    administration, 51, 52, 115, 117, 119, 167
    cavity, 121, 122
    delivery, 16, 17, 58, 96, 131
Organic
    inorganic hybrid film, 80
    solvents, 78, 94, 142, 169
Orthopedic materials, 150

Osmolarity, 37
Osteoblast phenotype, 154
Osteomyelitis condition, 101
Ovalbumin, 17, 131

P

Papain, 2
Papilloma, 22
Paracetamol (PM), 99, 120, 121
Peptide delivery systems, 38, 59
Peptides, 46, 57
PH
 dependent solubility, 167
 measurement, 21
 responsive hydrogel system, 16
 sensitive agents, 47
Pharmaceutical, 1, 2, 5, 8, 31, 37–39,
 45–47, 49, 57, 59, 71, 72, 74, 80, 82,
 84, 85, 87, 88, 96, 106, 111–113, 119,
 120, 122, 132, 141, 142, 145, 146, 150,
 155, 157, 165–167, 169, 170, 173, 181,
 182, 186
 agar, 74
 applications, 1, 31, 38, 57, 71, 72, 82,
  84, 85, 96, 111, 112, 132, 141, 155,
  157, 165–167, 181, 182, 186
  capsules/tablets for oral drug delivery,
   47
  gastroretentive drug delivery systems,
   54
  gelling agents, 46
  nanoparticles for drug delivery, 51
  oral particulates drug delivery, 49
  protein delivery, 57
  thickening/emulsifying agents/stabi-
   lizing, 46
 diluent, 47
 dosages, 37
 efficiency, 141
 excipients, 37, 38, 46, 57, 59, 85
 industry, 71, 74, 80, 84, 87, 146
 products, 37, 39, 46, 120
 research, 72, 170
Phosphate buffer
 saline, 132
 solution, 10, 17, 24

Photo-crosslinked methacrylated gellan
 gum (phGG-MA), 92
Photosensitivity, 104
Photothermal therapy, 71
Photo-ultraviolet irradiated, 28
Phycocolloid, 73, 74
Physicochemical
 characterization, 114
 properties, 88, 99, 112, 122, 166, 170,
  179, 180, 183, 186
PluronicF-68, 9
Poly(L-lactide-co-glycolide) (PLGA), 58,
 101
Poly(N-isopropylacrylamide), 101, 155
Poly(vinyl alcohol), 53, 112
Polyacrylamide, 50, 123, 183
Polyacrylic acid, 116, 123, 126
Poly-cationic polymers, 180
Polydispersity index, 13, 26
Polyelectrolytes, 130
Polyethylene glycol succinate, 27
Polyethyleneimine (PEI), 105
Polygalactan, 2
Polyguluronate
 residues, 43
 segments, 43
Polylactic glycolic acid, 10
Polymer
 chain, 89, 91, 92, 101, 176
 concentration, 98, 99
 matrix, 102, 178, 182
 network, 99, 123, 155, 178, 182
Polymeric
 materials, 72, 178
 segments, 42, 43
 solution, 98
Polymorphic forms, 37
Poly-morpho nuclear neutrophils cells, 125
Polysaccharide, 1, 2, 31, 38, 39, 53, 57,
 59, 72–75, 77, 82, 84, 87–89, 91, 95, 96,
 102, 106, 111–115, 118, 119, 122–124,
 126, 130–132, 141–143, 157, 165–169
Polyvinyl
 alcohol (PVA), 83, 105, 125, 155, 172,
  178, 182
 pyrrolidone (PVP), 83, 122
Povidone-iodine, 15, 122

Prednisolone, 52, 101, 115
Probiotic bacteria, 20, 21, 31
Prostaglandins, 114
Protamine, 131
Proteins, 39, 43, 46, 57, 58, 76, 90, 111, 112, 120, 131, 132, 143
Pseudomonas aeruginosa, 144
Pterocladia, 74

Q

Quercetin, 11, 12

R

Radical scavenging effects, 18
Regenerative medicine, 87, 125
Respiratory tract, 120
Rhamnose, 90, 94, 96
Rheological
  characteristics, 41
  properties, 102, 103, 116, 166, 184
Rhodophyceae, 2, 74
Rifampicin, 15

S

Salbutamol
  drug, 82
  sulfate, 50, 54, 57
Salicylazosulfapyridine, 115
Scanning electron microscopy, 7, 28, 123
Selenomethionine, 28, 29
Sericin, 6, 7
Serratia marcescens, 144
Sexually transmitted disease, 21, 22
Silver nanoparticles, 80, 174
Simulated
  gastric fluid (SGF), 21, 98, 118
  intestinal fluids (SIF), 98
Sinusitis operation, 152
Sodium
  acrylate, 3, 4
  alginate, 3, 5, 7, 9–11, 13–17, 41–43, 45, 47, 48, 50, 53–55, 57–59, 84, 123
    polylactic-co-glycolic acid, 10
    solutions, 41, 42
  cyanoborohydride, 148
  hydroxide, 23, 144

tripolyphosphate, 26–29
Sol-gel transformations, 42
Solid orals, 98
Somatology, 87
Sphingomonas, 87, 88
  elodea, 88, 94
Sprague–dawley rats, 119
Staphylococcus spp, 101
  S. aureus, 10
Succinic anhydride, 11, 27
Sulforhodamine, 23
Sulfuric acid, 76, 144
Syneresis method, 76

T

Tablet formulations, 111, 116, 118, 176
Testosterone
  injection, 51
  loaded nanocapsules, 51
Tetrahydrate, 14
Tetramethyl
  ethylenediamine, 3, 13
  orthosilicate, 5
Tetra-saccharide, 87, 88
Theophylline releasing
  capability, 48
  characteristics, 47
Therapeutic
  activity, 96
  agents, 115, 119
  properties, 132
Thermal
  hyperalgesia model, 115
  hysteresis, 78
  stability, 96, 169
Thermo-responsive
  microparticles, 101
  volume transitions, 131
Thermo-reversible gel, 78
Thermosensitive properties, 101
Thermostability, 150
Thioglycolic acid, 93, 101
Thiolation, 93
Thiol-bearing moiety, 147
Thrombogenic activity, 150
Thyme oil, 16

Tissue engineering, 87, 102, 105, 106, 112, 125, 127, 128, 132, 133, 141, 142, 149, 153–157
Toxic substances, 80
Toxicity, 15, 17, 23, 27–29, 31, 88, 131, 132
Traditional method, 75, 76
Tranexamic acid, 5, 6
Transdermal drug delivery devices, 156
Trans-mucosal delivery, 100
Trastuzumab, 27
Trifluoroethanol, 10
Trimethyl chitosans, 147
Trypanocidal, 1, 28, 31
Trypanosoma evansi, 27

U

Ulcerative colitis conditions, 115
United States Food and Drug Administration Ordinance (USFDA), 72, 166, 169

V

Vaccine delivery, 112, 120
Vanillin, 7, 8
Vascular endothelial growth factor, 113
Vasculature, 128
Vero cells, 18
    cell line, 27
Vessel-like networks, 128
Viscoelasticity, 117, 126
Viscosity, 5, 8, 41, 42, 48, 50, 59, 79, 88, 94, 95, 116, 117, 119, 120, 123, 167, 173, 174, 181, 183, 184

W

Water
    bioactive materials, 149
    molecules, 3, 78, 173
    soluble
        impurities, 76
        polymers, 148
Welan gum, 88
Wound healing, 1, 10, 15, 31, 46, 83, 104, 106, 112, 122, 132, 133, 141, 142, 152, 153, 156, 157

X

Xanthan
    gum (XG), 88, 118, 120, 165–186
    microspheres, 156
Xanthomonadaceae, 168
Xanthomonas campestris, 165–168
Xantural, 168
X-ray
    contrast media, 183
    scattering, 21

Z

Zaltoprofen, 18
Zein coat, 28
Zero-order model, 118
Zeta potential, 4, 5, 12, 15, 28, 29
Zidovudine, 8, 9
Zinc alginate, 45